ASSURING THE U.S. DEPARTMENT OF DEFENSE A STRONG Science, Technology, Engineering, and Mathematics (STEM) Workforce

Committee on Science, Technology, Engineering, and Mathematics Workforce
Needs for the U.S. Department of Defense and the U.S. Defense Industrial Base

Division on Engineering and Physical Sciences
with
Board on Higher Education and Workforce
Division on Policy and Global Affairs

NATIONAL ACADEMY OF ENGINEERING *AND*
NATIONAL RESEARCH COUNCIL
OF THE NATIONAL ACADEMIES

THE NATIONAL ACADEMIES PRESS
Washington, D.C.
www.nap.edu

THE NATIONAL ACADEMIES PRESS 500 Fifth Street, NW Washington, DC 20001

NOTICE: The project that is the subject of this report was approved by the Governing Board of the National Research Council, whose members are drawn from the councils of the National Academy of Sciences, the National Academy of Engineering, and the Institute of Medicine. The members of the committee responsible for the report were chosen for their special competences and with regard for appropriate balance.

This study was supported by contract number HQ0034-10-D-0003, delivery order 0003, between the National Academy of Sciences and the U.S. Department of Defense. Any opinions, findings, conclusions, or recommendations expressed in this publication are those of the author(s) and do not necessarily reflect the views of the organizations or agencies that provided support for the project.

International Standard Book Number-13: 978-0-309-26213-2
International Standard Book Number-10: 0-309-26213-5

Additional copies of this report are available from the National Academies Press, 500 Fifth Street, NW, Keck 360, Washington, DC 20001; (800) 624-6242 or (202) 334-3313; http://www.nap.edu.

Printed in the United States of America

THE NATIONAL ACADEMIES

Advisers to the Nation on Science, Engineering, and Medicine

The **National Academy of Sciences** is a private, nonprofit, self-perpetuating society of distinguished scholars engaged in scientific and engineering research, dedicated to the furtherance of science and technology and to their use for the general welfare. Upon the authority of the charter granted to it by the Congress in 1863, the Academy has a mandate that requires it to advise the federal government on scientific and technical matters. Dr. Ralph J. Cicerone is president of the National Academy of Sciences.

The **National Academy of Engineering** was established in 1964, under the charter of the National Academy of Sciences, as a parallel organization of outstanding engineers. It is autonomous in its administration and in the selection of its members, sharing with the National Academy of Sciences the responsibility for advising the federal government. The National Academy of Engineering also sponsors engineering programs aimed at meeting national needs, encourages education and research, and recognizes the superior achievements of engineers. Dr. Charles M. Vest is president of the National Academy of Engineering.

The **Institute of Medicine** was established in 1970 by the National Academy of Sciences to secure the services of eminent members of appropriate professions in the examination of policy matters pertaining to the health of the public. The Institute acts under the responsibility given to the National Academy of Sciences by its congressional charter to be an adviser to the federal government and, upon its own initiative, to identify issues of medical care, research, and education. Dr. Harvey V. Fineberg is president of the Institute of Medicine.

The **National Research Council** was organized by the National Academy of Sciences in 1916 to associate the broad community of science and technology with the Academy's purposes of furthering knowledge and advising the federal government. Functioning in accordance with general policies determined by the Academy, the Council has become the principal operating agency of both the National Academy of Sciences and the National Academy of Engineering in providing services to the government, the public, and the scientific and engineering communities. The Council is administered jointly by both Academies and the Institute of Medicine. Dr. Ralph J. Cicerone and Dr. Charles M. Vest are chair and vice chair, respectively, of the National Research Council.

www.national-academies.org

COMMITTEE ON SCIENCE, TECHNOLOGY, ENGINEERING, AND MATHEMATICS WORKFORCE NEEDS FOR THE U.S. DEPARTMENT OF DEFENSE AND THE U.S. DEFENSE INDUSTRIAL BASE

NORMAN R. AUGUSTINE (NAS[1]/NAE[2]), *Co-chair*, Lockheed Martin Corporation (retired)
C.D. (DAN) MOTE, JR. (NAE), *Co-chair*, University of Maryland, College Park
BURT S. BARNOW, George Washington University
JAMES S.B. CHEW, L-3 Communications
LAWRENCE J. DELANEY, Titan Corporation (retired)
MARY L. GOOD (NAE), University of Arkansas at Little Rock
DANIEL E. HASTINGS, Massachusetts Institute of Technology
ROBERT J. HERMANN (NAE), Private Consultant, Bloomfield, Connecticut
J.C. HERZ, Batchtags, LLC
RAY O. JOHNSON, Lockheed Martin Corporation
ANITA K. JONES (NAE), University of Virginia
SHARON LEVIN, University of Missouri-St. Louis
FRANCES S. LIGLER (NAE), Naval Research Laboratory
AARON LINDENBERG, Stanford University
PAUL D. NIELSEN (NAE), Software Engineering Institute, Carnegie Mellon University
DANIEL T. OLIVER, Naval Postgraduate School
C. KUMAR N. PATEL (NAS/NAE), Pranalytica, Inc.
LEIF E. PETERSON, Advanced HR Concepts and Solutions, LLC
STEPHEN M. ROBINSON (NAE), University of Wisconsin-Madison
MICHAEL S. TEITELBAUM, Harvard Law School
RONALD WILLIAMS, The College Board

Staff

TERRY JAGGERS, Lead Board Director
MARTIN OFFUTT, Study Director
CATHERINE DIDION, Senior Program Officer
GAIL GREENFIELD, Senior Program Officer
DANIEL E.J. TALMAGE, JR., Program Officer
KAMARA BROWN, Research Associate (through January 2012)
SARAH CAPOTE, Research Associate
MARGUERITE SCHNEIDER, Administrative Coordinator
DIONNA ALI, Senior Program Assistant

[1] NAS = member, National Academy of Sciences.
[2] NAE = member, National Academy of Engineering.

Preface

This report on the science, technology, mathematics, and engineering (STEM) workforce of the Department of Defense (DOD) and the U.S. defense industrial base marks the conclusion of an 18-month study to assess the STEM capabilities that the DOD will need in order to meet its responsibilities and priorities; to assess whether the current DOD workforce and personnel strategies will meet those needs; and to identify and evaluate options and recommend strategies that the department could use to enhance its effectiveness in meeting its future STEM needs. The study was undertaken jointly by the National Academy of Engineering and the National Research Council at the request of the Honorable Zachary J. Lemnios, Assistant Secretary of Defense for Research and Engineering (ASD[R&E]).

The committee preparing this report, the Committee on Science, Technology, Engineering, and Mathematics Workforce Needs for the U.S. Department of Defense and the U.S. Defense Industrial Base, initially convened a workshop on August 1 and 2, 2011, in Rosslyn, Virginia, for the purpose of gathering a broad range of views from the public sector and the private sector, including major defense contractors, and from nongovernmental organizations (NGOs), all of which are stakeholders in the future STEM workforce. A report issued in early 2012 summarized the views expressed by individual workshop participants.[1] An interim report was issued in June 2012 for the purpose of assisting ASD(R&E) with its fiscal year (FY) 2014 planning process and with laying the groundwork for future years.[2]

The present report highlights and addresses the critical need for scientists and engineers within DOD and its contractors, the latter to the extent they are engaged in defense-related activities.

---

[1]National Research Council. 2012. *Report of a Workshop on Science, Technology, Engineering, and Mathematics (STEM) Workforce Needs for the U.S. Department of Defense and the U.S. Defense Industrial Base.* Washington, D.C.: The National Academies Press.

[2]National Research Council. 2012. *An Interim Report on Assuring DOD a Strong Science, Technology, Engineering, and Mathematics (STEM) Workforce.* Washington, D.C.: The National Academies Press.

CAVEAT

It is emphasized that this report does not examine fulfilling the nation's overall demand for science and engineering talent. Indeed, important differences exist between defense and commercial needs in these fields, not the least of which is the result of the steep decline projected for defense spending, which portends a reduction in opportunities for most, but not all, categories of engineering and scientific talent within the defense sector. The foreseeable consequence for defense is primarily the need to assure the high quality of the workforce as opposed to its quantity.

Most commercial activities have become sufficiently internationalized and globalized that the STEM talent base is itself a global pool. Under these circumstances, the demand for scientists and engineers physically based in the United States often does not, per se, drive personnel decision making. Rather, in this instance, the issue becomes a national one of whether the jobs created through the efforts of scientists and engineers are located in the United States or elsewhere. The latter question, although critically important to the nation as a whole, was not a subject of this report.

ACKNOWLEDGMENTS

We wish to express our appreciation to the members of the committee for their diligent and dedicated contributions to the study and to the preparation of this report. The committee's diverse experiences contributed greatly to the broad perspective on STEM workforce evident in this report. We also wish to thank Stephanie Brown of the Naval Postgraduate School for her dedicated attention to the committee's discussions and its preparation of Chapter 5. The committee cannot thank the NRC staff members, Terry Jaggers, Martin Offutt, Gail Greenfield, Daniel E.J. Talmage, Jr., Kamara Brown, Sarah Capote, Marguerite Schneider, and Dionna Ali, and NAE staff member Catherine Didion, too effusively for their dedication to the study and to the preparation of this report.

Norman R. Augustine, *Co-chair*
C.D. (Dan) Mote, Jr., *Co-chair*
Committee on Science, Technology, Engineering, and Mathematics Workforce Needs for the U.S. Department of Defense and the U.S. Defense Industrial Base

Acknowledgment of Reviewers

This report has been reviewed in draft form by individuals chosen for their diverse perspectives and technical expertise, in accordance with procedures approved by the National Research Council's (NRC's) Report Review Committee. The purpose of this independent review is to provide candid and critical comments that will assist the institution in making its published report as sound as possible and to ensure that the report meets institutional standards for objectivity, evidence, and responsiveness to the study charge. The review comments and draft manuscript remain confidential to protect the integrity of the deliberative process. We wish to thank the following individuals for their review of this report:

Wanda Austin, NAE, The Aerospace Corporation
Lynda Carlson, National Science Foundation (retired)
VADM (ret.) Paul Gaffney, NAE, Monmouth University
Maj. Gen. (ret.) Robert Latiff, Independent Consultant
Kaushik Rajashekara, NAE, University of Texas at Dallas
Steven Ramberg, National Defense University
Harold Salzman, Rutgers, the State University of New Jersey
John Sommerer, Johns Hopkins University Applied Physics Laboratory
Paula Stephan, Georgia State University
David Whelan, NAE, The Boeing Company

Although the reviewers listed above have provided many constructive comments and suggestions, they were not asked to endorse the conclusions or recommendations nor did they see the final draft of the report before its release. The review of this report was overseen by Lawrence D. Brown, NAS, University of Pennsylvania, and Martha A. Krebs, University of California, Davis. Appointed by the NRC, they were responsible for making certain that an independent examination of this report was carried out in accordance with institutional procedures and that all review comments were carefully considered. Responsibility for the final content of this report rests entirely with the authoring committee and the institution.

Contents

Figures, Tables, and Boxes

FIGURES

TABLES

BOXES

Summary

THE REALITY

America's ability to fund, and thereby accomplish, its national security goals depends heavily on the strength of the nation's economy. The vibrancy of that economy has in turn been shown to depend heavily on advancements in science and engineering (National Research Council, 2007). Similarly, the ability of the nation's military to prevail during future conflicts, particularly while minimizing casualties, and to fulfill its humanitarian and other missions depends heavily on continued advances in the nation's technology base. A workforce with robust science, technology, engineering and mathematics (STEM) capabilities is critical to sustaining U.S. preeminence.

Today, however, the activities of the Department of Defense (DOD) devoted to science, technology, engineering, and mathematics are a small and diminishing part of the nation's overall science and engineering enterprise. One consequence is that DOD cannot significantly impact the nation's overall STEM workforce—and therefore, with a few exceptions, DOD should focus its limited resources on fulfilling its own special requirements for STEM talent.

THE DILEMMA

As a general rule, a student must decide in the 8th grade or earlier whether to preserve the option to pursue a career in STEM fields because of the hierarchical learning of mathematics (the "language" of STEM). In the traditional U.S. education course it takes about 8 more years for an individual in the 8th grade to graduate with a bachelor's degree in science or engineering—and about 14 more to graduate with a PhD in one of those fields.

Even setting aside the shortcomings of DOD's management of its STEM assets, the historical record of forecasting the number of scientists and engineers needed to work in national security has been abysmal at best, largely owing to inherent uncertainties in future threats and to the unpredictability of future technological advancements.

As to predicting military demands, history has proven that our best efforts cannot predict surprise events. World War I was triggered when an archduke was unexpectedly murdered and an unprepared America subsequently became entangled in conflict. U.S. involvement in World War II was sparked by the surprise attack on Pearl Harbor; in the Korean conflict, by a surprise assault across the 38th parallel; in Vietnam, by an unanticipated incident at sea; and in Afghanistan, by a surprise terrorist attack on U.S. soil. The current upheaval in the Middle East started with an altercation between a street vendor and a policeman.

Turning to technology as it applies to the military, the ability to forecast significant advancements has hardly improved between the invention of the riding stirrup and the discovery of stealth materials and shapes. Indeed, looking back 40 years—or even 10 years—few would have predicted the technology that is available today in either the military or the civilian spheres. Further, the pace of technological progress appears to be accelerating, not stabilizing or slowing.

The relatively small fraction of U.S. citizens graduating with first degrees in a STEM field (National Science Board, 2012, p. O-7), combined with our demonstrated inability to forecast sudden increases in demand for specialized STEM workers to support national security needs, can place the nation in jeopardy.

CHANGING FACTORS INFLUENCING THE DOD STEM WORKFORCE

Two fundamental changes—ironically, both are driven by advancements in science and engineering—have further complicated the above already complex situation. The first of these is the phenomenon described by Frances Cairncross: distance is dead (Cairncross, 1997). Indeed, globalization means that for many human endeavors distance is no longer significant, whether it is offshoring software development or attacking targets in Afghanistan using robots operated from Nevada. The second fundamental change is that for the first time in history individuals or small groups of individuals acting alone can profoundly impact the lives of very large groups of people.

But the revolutionary change now being experienced in both civilian and military affairs does not stop with these two groundbreaking developments. Other lesser but still profound changes affect DOD's need to recruit and retain high-quality scientific and engineering talent. These include:

- New technological opportunities and threats that are appearing with ever-increasing frequency (National Research Council, 2012b).
- The fact that for many technologies the most advanced work is no longer being conducted in the United States (National Research Council, 2006, 2010c; Naval Research Advisory Committee, 2010),
- The further fact that for most technologies, the most advanced work is no longer being conducted within the Department of Defense or its contractor community (Defense Science Board, 2012).
- The growing hazard to U.S. security posed by failed states (U.S. Department of Defense, 2010) .
- The erosion of the concept of deterrence based on possession of superior military weapons because of so-called asymmetric threats and, potentially, further nuclear proliferation (Drell, 2007; *Economist*, 2012).
- Inability to control knowledge because information penetrates porous geopolitical borders literally at the speed of light (National Research Council, 2006).
- Expansion of national security demands, with the real threat of conventional conflicts in places such as Korea, the Middle East, and possibly the Arctic and with the vastly different type of conflict introduced by terrorism (Jordan et al., 2009).

CURRENT OUTLOOK

The increasing importance of STEM in maintaining a strong economy and providing national security makes it imperative that America have available a substantial, high-quality STEM workforce. However, as compared with the young people of many other countries, American youth seem less interested in pursuing careers in STEM fields. In the recent past this development has been substantially offset by attracting foreign-born individuals to America's research universities and then making it possible for them to remain and contribute to America's well-being and to their own quality of life. Of the current science and engineering workforce outside academia, one-quarter are foreign born (National Science Board, 2012, p. 3-48).

Today, more than one-half the PhD's awarded by U.S. engineering schools go to non-U.S. citizens. Of those non-U.S. citizens who graduated with science and engineering doctorates in 2004, 38 percent had left the United States 5 years later (National Science Board, 2012, p. 3-51). The fraction of master's degrees awarded to temporary visa holders is smaller but increasing (Figure S-1). Bachelor's degree holders constitute half of DOD's STEM workforce, and non-U.S. citizens have consistently earned 3 to 4 percent of U.S.-awarded bachelor's degrees, although in certain fields, such as electrical and industrial engineering, the fraction is higher, at 9 percent (National Science Board, 2012, p. 2-22).

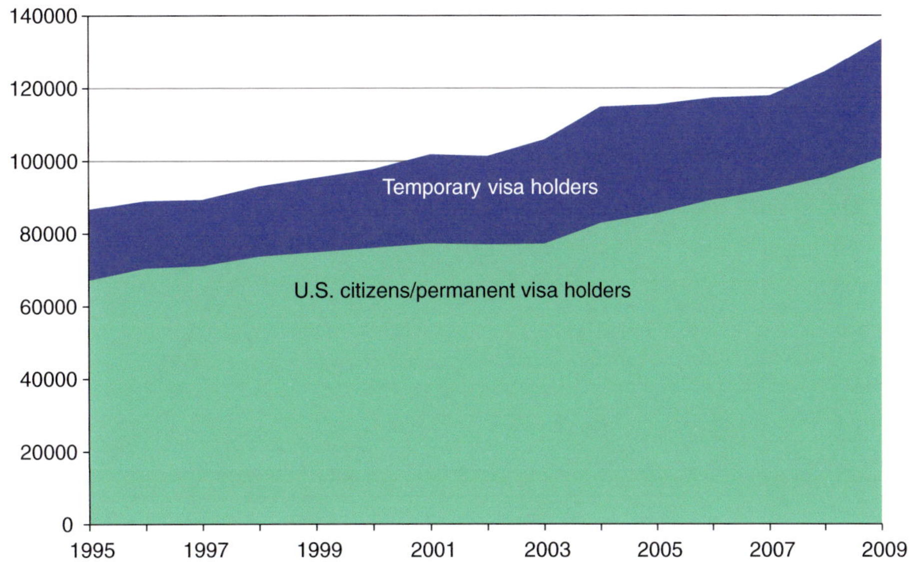

FIGURE S-1 Number of master's degrees awarded in the United States, by visa status.
SOURCE: Lehming (2011).

However, the process by which the United States met its workforce needs so well in the past is in jeopardy, for several reasons:

- U.S. national immigration policy places caps on the number of high-tech (i.e., H1-B) visas allotted to for-profit organizations, and this pool of visa holders is an important source of scientists and engineers, while the coveted green card conferring permanent work status can take 6 to 10 years to obtain. In the short run, further constraints on H1-B visa entrants may make it more difficult for DOD to recruit citizens if these constraints increase competition for them from the private sector.
- Individuals who manage to overcome the barriers posed by U.S. immigration laws and remain in the United States as noncitizens after receiving their degrees are excluded from most defense-related work because of the associated requirement to hold a security clearance and the rigidity of the security clearance process (National Research Council, 2010b).
- Opportunities are increasing in many parts of the world for scientists and engineers—both U.S. citizens and noncitizens—to build productive careers in other lands because talent is in such widespread demand (Wadhwa et al., 2009).
- The current DOD science and engineering workforce is an aging one (Figure S-2), with a disproportionate segment of scientists and engineers eligible to retire during the next few years (Figure S-3).
- Despite an increase in the percentage of the defense industrial base STEM workforce that is under the age of 35, the median age of such workers increased to 47 in 2010, from 45 in 2005.
- A recent survey of over 59,000 college students in various fields of study at over 300 universities assessed the desirability of potential employers. In engineering fields, the Air Force ranked 15th, followed by the Navy at 34th and the Army at 41st. In the natural sciences, the Air Force ranked 20th, followed by the Navy at 22nd and the Army at 25th. In neither of these two fields was DOD ranked in the top 100. In the field of information technology, however, DOD was ranked 20th, above the U.S Air Force at 31st, U.S. Navy at 34th, and U.S. Army at 60th (Universum, 2012).[1]

[1]In the survey, the interpretation of which organizational components were encompassed by "DOD," "U.S. Army," and so forth was left to the survey respondents.

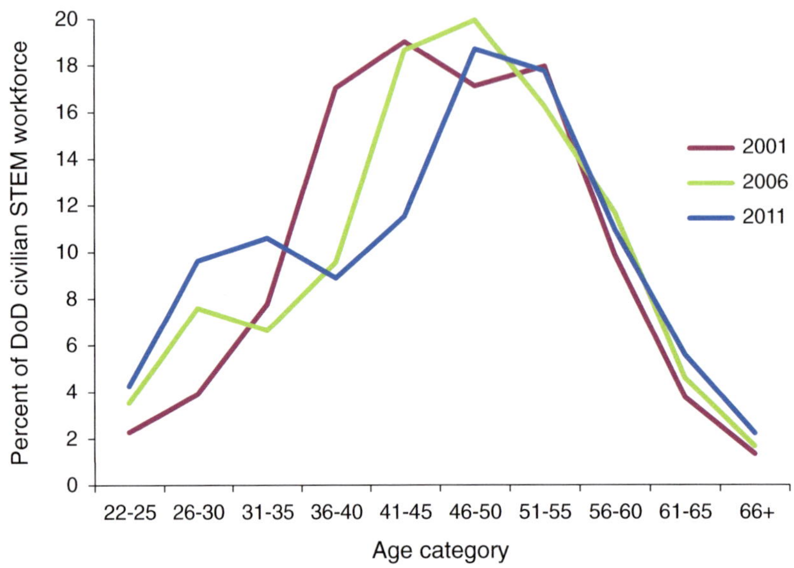

FIGURE S-2 Age distribution of Department of Defense civilian STEM workforce, selected years: 2001, 2006, and 2011.
NOTE: Figures are as of the fiscal year-end (e.g., September 30, 2011).
SOURCE: Data provided by the Defense Manpower Data Center. Tabulations by the National Research Council.

- The "defense industry," composed of the principal DOD contractors, is moving to diversify away from defense for economic reasons (Thompson, 2011)—and because of the complexities in dealing with a powerful monopsonist (i.e., a sole) buyer.
- Because of economic circumstances, the nation is unlikely to be able to support defense expenditures at the levels of the past (Appelbaum, 2012), and DOD's traditional predilection is not to give highest priority to funding for research (National Research Council, 2008, 2011).
- Technology today has a half-life measured in a few years, whereas major DOD development programs can take decades—making it nearly impossible under current practices to supply U.S. armed forces with the most advanced technology (National Research Council, 2010a, 2012a).

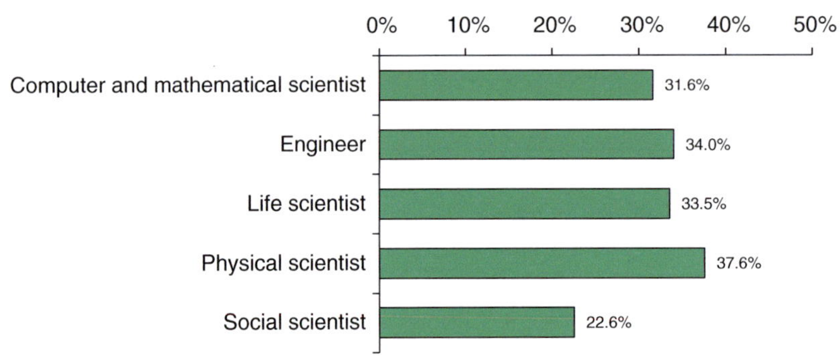

FIGURE S-3 Retirement eligibility of selected occupational groups in the DOD civilian STEM workforce.
NOTE: Percentages are as of the fiscal year-end (September 30, 2011).
SOURCE: Data provided by the Defense Manpower Data Center. Tabulations by the National Research Council.

• U.S. industry as a whole is further reducing its investment in research,[2] with, for example, iconic institutions such as Bell Labs now diminishing in size and no longer U.S. owned.

• Government contractors have become increasingly risk-averse, constrained as they are by increasingly complex defense acquisition laws (Dunlap, 2011) and competing for fewer acquisition programs that have longer acquisition cycles—all of which make the work less attractive to prospective STEM hires (National Research Council, 2012a).

• The U.S. higher education system finds its predominant global position threatened by declining investments in education by state and local governments as well as by greatly increasing competition from government-funded universities and research institutions abroad.

• The United States scores average or below average among OECD countries in the proficiency of its K-12 students (OECD, 2010), and U.S. nationwide testing has shown that the average 4th grader was less than proficient in mathematics and science.[3]

THE CONUNDRUM

U.S. employers nearly unanimously cite the need for additional employees with specialty skills, including STEM workers, yet the nation's overall unemployment rate remains high. Steve Jobs told the President that one of the reasons his firm had to employ 700,000 workers abroad was the ability of China to supply engineers much more rapidly than the United States, including 8,700 industrial engineers to oversee the 200,000 assembly-line workers, who were found in China in just 15 days (Duhigg and Bradsher, 2012; Wingfield, 2012). But what the United States confronts as a nation, and what DOD confronts to an even greater extent, is not an unemployment problem but a knowledge gap (i.e., a quality) problem, particularly with the potential STEM workforce.

DOD representatives state virtually unanimously that they foresee no shortage of STEM workers in the years ahead except in a few specialty fields such as cybersecurity and intelligence. However, the aerospace and defense industry has experienced difficulty in hiring systems engineers, aerospace engineers, and mechanical engineers. Pondering the projected decline in defense spending, it is not difficult to imagine a reduction in the perceived need for STEM employees by DOD and its contractors. The problem is that with the rapid pace of advancement in STEM and the uncertainty of future threats, a shortage of STEM workers, particularly those with knowledge in evolving fields, could occur at any time.

The DOD's STEM needs, as well as those of its contractors, represent a relatively modest facet of the challenge faced by the nation's workforce as a whole in today's burgeoning, technologically driven economy. Total DOD civilian STEM employment is approximately 150,000, with 47 percent in engineering and 35 percent in computer and mathematical science occupations; this workforce represents only a small fraction (approximately 2 percent) of the total U.S. STEM workforce. For the private sector, although STEM jobs are a major component of the defense industrial base (approximately 3 in 10 jobs), these jobs also represent a small fraction of total U.S. STEM employment (likewise approximately 2 percent). A notable exception is aerospace engineers, a substantial proportion of whom are employed in the aerospace and defense industry.

Ironically, it is unlikely that the United States will suffer from an overall shortage of scientists and engineers. The principal reason is globalization. Today, it is a relatively straightforward matter for a U.S. commercial firm to fulfill its STEM capacity needs abroad—particularly given the large numbers of STEM workers being educated elsewhere in the world, a growing number of whom are highly qualified.

As U.S. industry's research laboratories move abroad (National Science Board, 2012, Figures O-6 and O-7), so too do the prototype shops that design and evaluate new concepts, and so too do the production lines and eventually the maintenance facilities (in order to reap higher returns on their investment (*Economist*, 2011))—and so too do the continuous design modifications over the product life cycle and the ideas for subsequent innovations and generation of equipment. Further, most of tomorrow's commercial customers will be in the developing nations,

[2]The R&D investment by U.S. business declined faster than GDP in 2008-2009 and the decade ending in 2009 saw a slowing of R&D expenditures versus earlier periods. See for example, Chapter 4 in National Science Board (2012).

[3]See, for example, Figures 8-1 and 8-4 in National Science Board (2012).

not in the developed countries as in the past,[4] making it all the more attractive to conduct manufacturing and engineering outside the United States. A principal outcome of this scenario is that there will not be enough jobs in the United States for U.S. workers as a whole, and unemployment will remain high.

Another complication related to the security of our nation is that DOD and its contractors cannot simply export their work to overseas firms—although DOD will need to do a much better job of defining exactly which jobs truly demand U.S. citizenship as a condition of employment. The maintenance of a cadre of highly capable, dedicated, innovative, entrepreneurial U.S. scientists and engineers is thus critical to the health of the U.S. economy as well as that of DOD.

In this context, DOD's demand for scientists and engineers is sufficiently modest that fulfilling its need for *numbers* should be achievable. DOD's challenge in the foreseeable future is filling its ranks with a suitable share of the best and brightest talent—particularly given the current perception of many young graduates, in particular PhD candidates in the sciences, that working in government is less compelling, though still attractive, than careers in academic teaching and research or industry (Sauermann and Roach, 2012).

The highly regarded Science, Mathematics and Research for Transformation (SMART) Scholarship for Service Program is a DOD STEM workforce development program that addresses recruiting and retaining top talent for the department. It is a civilian scholarship-for-service program that provides full undergraduate or graduate tuition, living and book allowances, summer internships, health insurance, and other benefits in exchange for postgraduate employment at DOD; the scholarship is paid back by service on a one-year-for-one-year basis. The qualification of the students is high—the 2009 cohort of 262 students had a GPA of 3.7. This 6-year program is attractive, expandable, and well-targeted to the nation's national security needs.

There are a number of constructive goals DOD could set to help assure that the needed cadre of highly qualified STEM workers will be available to support U.S. national security needs. These include (1) making the DOD a more attractive place for highly capable STEM employees to work; (2) creating more pathways for high-quality scientists and engineers to work in DOD; (3) enhancing early warning of new developments being achieved globally in science and engineering by increasing the involvement of DOD's workforce in global activities in core fields; (4) managing the careers of high-quality civilian government scientists and engineers and giving them educational opportunities, as is already done for the most capable uniformed personnel; and (5) establishing and ensuring adaptable human resource development and management mechanisms that can respond to abrupt changes in STEM opportunities and needs that are fully competitive with the responsiveness found in industry.

PRINCIPAL FINDINGS

Science and technology and the DOD STEM workforce are increasingly critical to U.S. military capability. Technological surprise has proved to be decisive in past conflicts and will likely be so in the future. The ongoing globalization of STEM requires that DOD readdress its workforce policies and practices to ensure that it retains access to a significant share of the best and brightest STEM talent available. DOD is a microcosm of the larger and growing global STEM enterprise, where talent is in high demand. Access to highly qualified STEM talent should be a primary consideration in DOD workforce recruitment and retention policies, guidelines, and practices.

Finding 1: Quantity of STEM Workforce

Because of the relatively small and declining size of the DOD STEM workforce there is no current or projected shortage of STEM workers for DOD and its industrial contractor base except in specialized, but important, areas—such as cybersecurity and selected intelligence fields. As a means of addressing any future shortages, experience has shown that students will respond to the demand signal of higher salaries in a STEM field[5] (Figure S-4), sug-

[4]Asia's spending on defense is projected to surpass that of Europe in 2012. For more information see International Institute for Strategic Studies (2012).

[5]The committee was made aware of a further instance in which students' choice of a STEM major was made in response to the offer of higher salaries, though it was for the case of petroleum engineers, a field for which DOD has little if any need. See NRC (2012a), p. 26.

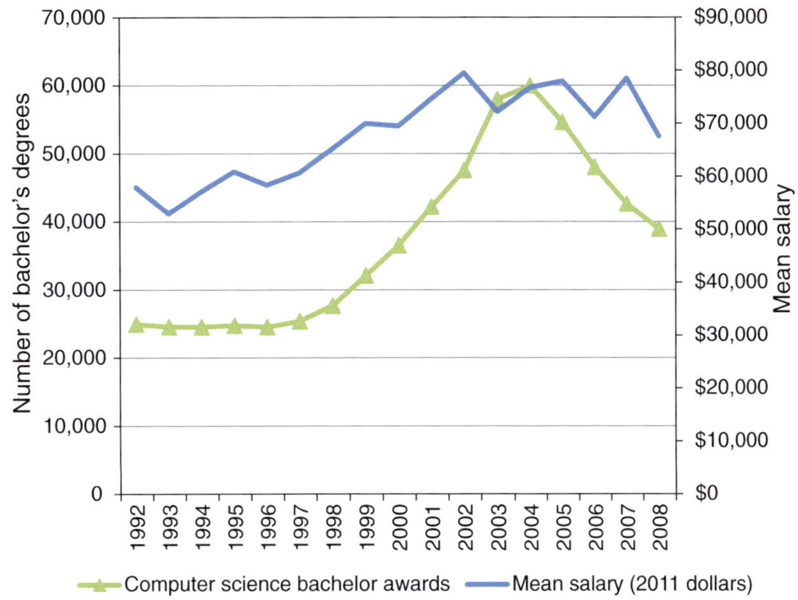

FIGURE S-4 Computer science bachelor's degree awards and computer programmer real mean salaries, 1992-2008.
SOURCE: Kuehn and Salzman (2013).

gesting a mechanism by which DOD can stimulate supply in a critical area.[6] (See Observation 3-10, Observation 3-4, and Finding 2-5.)

Finding 2: Quality of STEM Workforce

The STEM issue for DOD is the quality of its workforce, not the quantity available. The DOD needs a suitable share of the most talented STEM professionals. The decisions they make within DOD are highly leveraged, impacting the efforts of very large numbers of people and enterprises both inside and outside the government. (See Finding 6-3.)

Finding 3: Changing Character of STEM Workforce

New technological advancements, often from outside the defense sector and from abroad, are appearing at an increasing rate. Adapting to this new environment requires transformational and long-term changes within the DOD management of its STEM workforce. (See Finding 6-1.)

Finding 4: Forecasting STEM Workforce Needs

Reliable forecasting of the STEM skills needed by the DOD beyond the near term is simply not possible because of the increasing rates of advancement in science and technology and the unpredictability of military needs. *Flexibility*, *capability*, and *relevance* in the DOD STEM workforce are the essential characteristics sought. (See Finding 6-6.)

[6]Freeman (1976) established that "the supply of new entrants to engineering is highly responsive to economic conditions."

Finding 5: Attracting and Retaining STEM Workforce

For DOD to recruit top STEM talent in competition with commercial firms, universities, and others, it must commit to improving the STEM workforce environment. The DOD must become, and be perceived as, an attractive career destination for the most capable scientists, engineers, and technicians who are in great demand in the global talent marketplace. This implies, among other things, that DOD will need to reassess its requirement for security clearances for many STEM positions along with the processes by which many of its systems are developed and procured. (See Finding 4-2 and Finding 4-3.)

Finding 6: Managing the STEM Workforce

The career development support for the DOD uniformed STEM workforce is excellent, whereas the career development support for the DOD civilian STEM workforce is far less developed. The defense-related industry lies somewhere between them. (See Finding 6-4.)

PRINCIPAL RECOMMENDATIONS

Based on the above findings, the study committee developed five principal recommendations. These are summarized in brief in the list that follows, with the suggested implementations described in the relevant chapters of the report.

The committee observes that the foreseeable STEM personnel challenge is, with the exception of a very few highly specialized disciplines, not one of meeting quantitative needs but one of providing the high-quality STEM personnel needed to fulfill the DOD mission at a high technical standard.[7] Because of the leadership role that DOD STEM personnel often play in overseeing major programs and directing the efforts of large groups within the private sector as well as impacting others in government, the STEM capability and quality of the DOD leadership in its workforce are highly leveraged.

Through focused investments DOD should ensure that STEM competencies in all potentially critical, emerging topical areas are maintained at least at a basic level within the department and its industrial and university bases. This appropach will ensure that technological challenges and opportunities that arise can be met expeditiously by building on the foundation that is in place.

Recommendation 1. Recruitment and Retention of Highest-Quality STEM Workforce

The DOD workforce recruitment policies and practices should be reviewed and overhauled as necessary to ensure that DOD is fully competitive with industry (not simply the "defense industry") in recruiting the highest-quality STEM talent. DOD should judge its recruiting competitiveness by the quality of its STEM hires, and it should continue to adjust its policies and practices until it has become fully competitive with overall industry and academia in the quality of its recruitments. (See Box S-1.) Such practices might include the following:

- More active outreach and recruitment efforts aimed at civilian hires of needed scientists and engineers that emphasize the many exciting technologies that are being developed by DOD and their potential contribution to the nation;
- New measures to expedite recruitment offers for occupations in which DOD determines that it must compete with more nimble corporate recruitment practices;
- Additional authority to expedite security clearances needed for such positions, including authority for temporary hiring into non-sensitive roles pending confirmation of security clearance; and

[7] The committee considered how "quality" might be defined or what metrics might be constructed to better track the quality of the workforce. The committee decided, however, that quality measures vary from one discipline to the next, making it infeasible to provide one overarching definition. Those with hiring authority will be in the best position to consider a job candidate's knowledge, skills, and abilities and to weigh the degree of significance of individual records of achievements and capabilities compared to those of others.

BOX S-1
Innovative Recruitment Policies and Practices at the Advanced Research Projects Agency-Energy (ARPA-E) and at the Naval Research Laboratory

The Advanced Research Projects Agency-Energy (ARPA-E) funds specific high-risk, potentially high-payoff energy research and development projects. ARPA-E has been set up to be a lean and agile organization with special hiring authority to bring on program directors and other program leadership with the ability to offer limited-term rotational assignments. Thus, individuals from all sectors are able to assume temporary positions lasting roughly 3 years. The agency empowers them to make technical and programmatic decisions for the projects they oversee.[1]

The Naval Research Laboratory (NRL) has recently added a direct hiring facility, the Distinguished Scholastic Achievement Appointment (DSAA), aimed at speeding the recruitment of entry-level candidates. This complements its existing direct-hire authority for persons holding advanced degrees in science and engineering. Under DSAA, managers have the opportunity to expedite hiring of candidates with an exceptional grade point average and are allowed to hire individuals based solely on their education. Candidates for certain job classifications and occupational series who possess a GPA of 3.5 or higher may be appointed without NRL having to advertise each position individually. The individual must hold a bachelor's, master's, or higher degree in the field of the position being filled. Managers may name/request a candidate from the list forwarded by the human resources office for one of the advertised positions.

[1]Based on Yehle (2011) and President's Council of Advisors on Science and Technology (2010).

- Actions to protect or "ring-fence" science and engineering positions determined by DOD to be critical capabilities, thereby protecting the loss of such capabilities due to RIFs and hiring freezes.

Further, the DOD STEM workforce management should have as a *primary objective* retaining its highest-quality talent. Talented individuals include STEM professionals ranging from technicians to systems engineers to the most advanced scientists and engineers working in specialty fields. It is critical to include those at the forefront of emerging, potentially critical technical areas, and those capable of moving rapidly into these new areas. The DOD must ensure that its STEM workforce management policies, procedures, and incentives (in short, its *business model*) achieve that outcome. Its business model should explicitly make careers in DOD attractive to top STEM talent. Achievement of this goal will require explicit support, commitment, and action by the highest level of DOD leadership. (See Recommendation 4.3 and Recommendation 4.6.)

Recommendation 2. Open More of the STEM Workforce Pool to non-U.S. Citizens

Because DOD and its contractors need access to the most talented STEM professionals globally, DOD should reexamine the need for security clearances in selected positions in order to permit non-U.S. citizens to enter the STEM talent pool available to DOD under tailored circumstances consistent with applicable law and regulation governing military goods and services and their export and deemed export. (See Box S-2.) Further, the H1-B visa system should be modified to provide the nation and DOD with a substantially larger pool of extraordinary talent in areas of need. (See Recommendation 4.2.)

BOX S-2
Recruitment of Non-U.S. Citizens at the National Laboratories

Sandia National Laboratories has a hiring pathway by which a foreign national can become a member of its technical staff. The first stage for such an individual is to become established as a staff member (e.g., in a postdoctoral position or as a limited-term employee). In the next stage the individual is given status as in a Foreign National Interim Technical Staff member, which includes a requirement that he/she concurrently pursue U.S. citizenship. Owing to the classified nature of the lab's work, the prospective staff member must obtain the necessary security clearances and successfully pass a comprehensive counterintelligence investigation. At this point, or upon receipt of citizenship, the individual becomes a member of the technical staff.

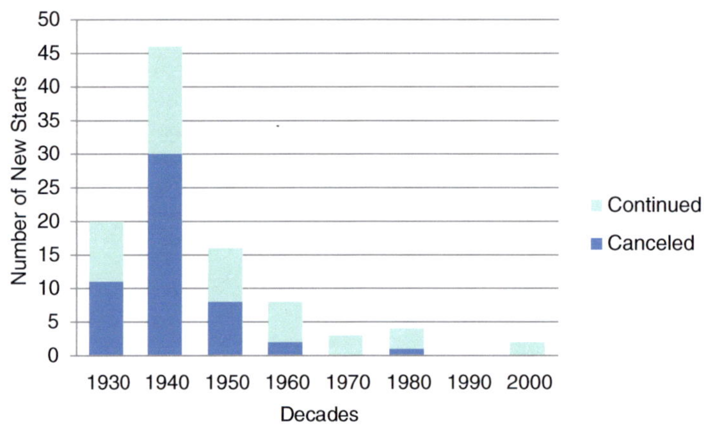

FIGURE S-5 Number of new fighter and bomber starts per decade.
SOURCE: Carlson and Chambal (2008).

Recommendation 3. Maintain Critical STEM Capabilities Through Unconventional Programs and Prototyping

To preserve design, creation, and testing team skills (which have been called on less and less as new weapons systems appear with decreased frequency—Figure S-5) and to recruit, retain, and advance a quality STEM workforce with the special talents needed by DOD and its contractors, DOD should create "skunk works"[8] in the industrial base, universities, and DOD to undertake targeted, unconventional, potentially disruptive programs through prototyping for technical concept verification. These programs could subsequently be transitioned to an operating unit for implementation if successful, or terminated if not. A system that provides rotational assignments for individuals from government, the industrial base, and the private sector would be an attractive feature of these programs. This "skunk works" culture would nurture critical STEM skills within the DOD workforce as well as provide exciting, challenging, and highly attractive opportunities for the STEM workforce. (See Box S-3.) (See Recommendation 4.4.)

[8]"Skunk works" refers to Lockheed Martin's Advanced Development Program for manned and unmanned systems, which began operations in the 1940s and has since designed numerous aircraft such as the U-2, the SR-71 and the F-111.

BOX S-3
Rapid Prototyping in the Office of the Assistant Secretary of Defense for Research and Engineering

DOD established the Rapid Reaction Technology Office (RRTO) in 2006 in response to the constantly evolving threat of asymmetric warfare, including, for example, the use of improvised explosive devices (IEDs) in the Iraq and Afghanistan theater of operations. Established under the Director, Defense Research and Engineering, the office focused on developing technologies that can mature in 6 to 18 months for the purpose of countering insurgency and irregular warfare. It now has been folded into the Rapid Fielding Office within ASDR&E. The RRTO provides a diverse set of quick-response capabilities for counterterrorism while attempting to stimulate interagency coordination and cooperation. The office operates without a formal charter or governing document, and the director has much flexibility for carrying out the mission. Approximately 50 percent of the office's projects have resulted in fielded technologies, altered concepts of operation (CONOPS), or other concrete changes, including in larger systems. Such projects included the Persistent Threat Detection System for persistent ground surveillance through a tethered aerostat with an embedded camera; the Biometric Automated Toolset for screening personnel in mobile applications; and the SKOPE intelligence cell, a joint analytic effort with the National Geospatial Intelligence Agency, the U.S. Special Operations Command, and the U.S. Strategic Command.[1]

[1]Adapted from NRC (2009).

Recommendation 4. Develop an Agile and Resilient STEM Workforce

The DOD should recruit and develop an agile and resilient STEM workforce that is attuned to the dynamism and future uncertainty of technical needs; is prepared to adapt to those needs as they arise; and is enthusiastic about working in this challenging environment. (See Box S-4.) In addition, the DOD should be prepared to educate highly capable, but not yet STEM qualified, individuals rapidly into STEM-capable professionals with master's degrees in science and engineering in times of urgent need—as is done at the Naval Postgraduate School today. (See Box S-5.) (See Finding 5-2 and Recommendation 5-2.)

Recommendation 5. Upgrade Education and Training for the DOD Civilian STEM Workforce

The DOD should ensure that the education and training, and the re-education and re-training, opportunities for its civilian STEM workforce are both commensurate with similar opportunities afforded career military personnel and tailored to the needs of the civilian workforce. (See Box S-6.) (See Recommendation 5-2 and Finding 6-4.)

AREAS OF NEAR-TERM FOCUS

Although it is the conclusion of this committee that planning for future STEM needs should be geared to flexibility and versatility rather than forecasting, certain areas do have strong *near-term* interest with a potential for high impact on future DOD operations. STEM personnel will create, recognize, and exploit breakthrough discoveries, engineer prototypes and operational versions for military use, and integrate them into systems controlled by humans. The identification of those areas is based on a combination of apparent needs and high promise and is meant to illustrate implications for the STEM skills needed by DOD and the industrial base. A listing of them in alphabetical order is as follows:

BOX S-4
Agile and Adaptable Workforce Practices at NASA and at Lockheed Martin

NASA created its Engineering and Safety Center (NESC) in 2003 to provide an independent test, analysis, and assessment capability for NASA programs and projects. It operates independently of mission directorates and reports to the Office of the Chief Engineer. The NESC operates through technical discipline teams (TDTs), each led by an agency-recognized NASA tech fellow, who is an outstanding senior-level engineer or scientist with distinguished and sustained records of technical achievement. The fellows provide leadership and act as role models for NASA discipline engineering communities beyond the TDT; they are drawn not only from NASA but also from other federal agencies, industry, and universities. The TDTs are diverse teams and can provide robust, creative solutions to complex problems.

Over its nearly 70-year history, the Lockheed Martin Skunk Works® has created breakthrough technologies and landmark aircraft that continually redefine flight. Guided by the mantra "quick, quiet, and quality," the Skunk Works requires a flexible workforce capable of quickly forming and disbanding interdisciplinary project teams. To meet this need, the Skunk Works uses a matrix organization that minimizes paperwork and delays in moving people between teams. Core engineering groups maintain skill sets and tools to support their disciplines. Program managers draw their teams from these talent pools.

BOX S-5
Rapid Retraining into Technical Fields at the Naval Postgraduate School

The Naval Postgraduate School grants master's degrees in engineering to selected individuals who enter with liberal arts credentials. Between 2007 and 2011 over 4,000 resident students graduated from this program, of whom roughly 525 had non-technical backgrounds when they matriculated. The education is accomplished via an intense, year-round academic program that focuses on technical master's degrees in engineering and other STEM coursework in curricula ranging from 18 to 30 months depending on the discipline and credentials of the incoming student.

BOX S-6
Graduate Study Programs for Members of the Military

The Department of Defense manages and funds postgraduate education of its military. A military authorization (i.e., a job position) can be coded with a requirement for an advanced academic degree (AAD) (PhD or master's). Within the Air Force, for example, such a requirement provides the leverage either to get a quota at the Air Force Institute of Technology (AFIT) or find a qualified person to fill that authorization. The Air Force regulation that addresses military AADs (an example of more formal support) exists but is outdated and being revised (AFI 36-2302, dated July 11, 2001, "Professional Development [Advanced Academic Degrees and Professional Continuing Education])." There is no equivalent Air Force regulation for civilians, with each career field managing its own postgraduate needs according to its own policies, practices, and funding levels.

- Advanced robotics and autonomous systems;
- Intelligence collection;
- Cyber warfare (defensive and offensive);
- Human identification, marking, and tracking;
- Human-machine interactions on human terms;
- Means to detect and neutralize bio-threats;
- Means to negate improvised explosive devices (IEDs);
- Military applications of biosciences (systems biology, biosensors, etc.);
- Military applications of information sciences;
- Nanotechnology (for innovative materials and other applications); and
- System design and integration.

BIBLIOGRAPHY

Appelbaum, B. 2012. The next war: A shrinking military budget may take neighbors with it. New York Times, January 6.

Cairncross, F. 1997. The Death of Distance. Boston, Mass.: Harvard Business School Press.

Carlson, G., and S. Chambal. 2008. Senior leader perspective. Developmental planning: The key to future war-fighter capabilities. Air and Space Power Journal 22(1):5-8.

Defense Science Board. 2012. Basic Research: Report of the Defense Science Board Task Force on Basic Research. Available at http://www.acq.osd.mil/dsb/reports/BasicResearch.pdf (accessed April 4, 2012).

Drell, S.D. 2007. Nuclear Weapons, Scientists, and the Post-Cold War Challenge: Selected Papers on Arms Control. Hackensack, N..J: World Scientific Publishing.

Duhigg, C., and K. Bradsher. 2012. How the U.S. lost out on iPhone work. New York Times, January 21.

Dunlap, C.J., Jr. 2011. The military industrial complex. Daedalus 140(3):12.

Economist, The. 2011. China's economy and the WTO. The Economist, December 10.

Economist, The. 2012. Nuclear security—Threat multiplier. The Economist, May 31.

Freeman, R.B. 1976. A cobweb model of the supply and starting salary of new engineers. Industrial and Labor Relations Review 29(2):236-248.

International Institute for Strategic Studies (IISS). 2012. The Military Balance. London: IISS.

Jordan, A.A., W.J. Taylor, Jr., M.J. Meese, S.C. Nielsen, and J. Schlesinger. 2009. American National Security. Sixth Edition. Baltimore, Md.: Johns Hopkins University Press.

Kuehn, Daniel, and Harold Salzman. 2013. The labor market for new engineers. U.S. Engineers in the Global Economy. Richard Freeman and Harold Salzman (eds.). National Bureau of Economic Research, forthcoming.

Lehming, R. 2011. STEM Workforce Needs of the U.S. Department of Defense: Background Data. Presentation to the Workshop on Science, Technology, Engineering, and Mathematics (STEM) Workforce Needs for the U.S. Department of Defense and the U.S. Defense Industrial Base, Roslyn, Va., August 1.

National Research Council. 2006. Critical Technology Accessibility. Washington, D.C.: The National Academies Press.

National Research Council. 2007. Rising Above the Gathering Storm: Energizing and Employing America for a Brighter Economic Future. Washington, D.C.: The National Academies Press.

National Research Council. 2008. Pre-Milestone A and Early Phase Systems Engineering. Washington, D.C: The National Academies Press.

National Research Council. 2009. Experimentation and Rapid Prototyping in Support of Counterterrorism. Washington, D.C.: The National Academies Press.

National Research Council. 2010a. Avoiding Technology Surprise for Tomorrow's Warfighter—Symposium 2010. Washington, D.C.: The National Academies Press.

National Research Council. 2010b. Critical Code: Software Producibility for Defense. Washington, D.C.: The National Academies Press.

National Research Council. 2010c. S&T Strategies of Six Countries: Implications for the United States. Washington, D.C.: The National Academies Press.

National Research Council. 2011. Evaluation of U.S. Air Force Preacquisition Technology Development. Washington, D.C.: The National Academies Press.

National Research Council. 2012a. Report of a Workshop on Science, Technology, Engineering, and Mathematics (STEM) Workforce Needs for the U.S. Department of Defense and the U.S. Defense Industrial Base. Washington, D.C.: The National Academies Press.

National Research Council. 2012b. A View of Global S&T Based on Activities of the Board on Global Science and Technology. Washington, D.C: The National Academies Press.

National Science Board. 2012. Science and Engineering Indicators 2012. Arlington Va.: National Science Foundation.

Naval Research Advisory Committee. 2010. Status and Future of the Naval R&D Establishment. Available at www.nrac.navy.mil/docs/2010_Summer_Study_Report.pdf (accessed October 17, 2011).

OECD. 2010. PISA 2009 Results: What Students Know and Can Do. Paris: Organisation for Economic Cooperation and Development.

President's Council of Advisors on Science and Technology. 2010. Report to the President on Accelerating the Pace of Change in Energy Technologies Through an Integrated Federal Energy Policy.

Sauermann, H., and M. Roach. 2012. Science PhD career preferences: Levels, changes, and advisor encouragement. PLoS ONE 7(5):e36307.

Thompson, L. 2011. Defense Contractors Are Going to Go for the Civilian Market. Forbes 2012, April 5. Available at http://www.forbes.com/sites/lorenthompson/2011/11/08/market-conditions-pressure-defense-companies-to-diversify (accessed November 8, 2011).

Universum. 2012. America's Ideal Employers 2102. Availiable at http://www.universumglobal.com/IDEAL-Employer-Rankings/The-National-Editions/American-Student-Survey (accessed May 31, 2012).

U.S. Department of Defense. 2010. Quadrennial Defense Review Report. Washington D.C.: Government Printing Office.

Wadhwa, V., A. Saxenian, R. Freeman, G. Gereffi, and A. Salkever. 2009. America's Loss Is the World's Gain: America's New Immigrant Entrepreneurs. Kansas City, Mo.: Ewing Marion Kauffman Foundation.

Wingfield, N. 2012. Apple's job creation data spurs an economic debate. New York Times, March 4.

Yehle, Emily. 2011. No home run yet for ARPA-E, but chief says "motivated" team's on track. Greenwire, April 7.

1

Introduction

OVERVIEW

Following World War II, the U.S. national security strategy was to ensure technological superiority in all critical military capabilities. Superiority was achieved through commitments to fundamental research in science and engineering and to creating superior weapons systems. Staying ahead technologically required (1) a superior STEM workforce within DOD, its private sector contractors, and academe; (2) significant and continuous investment in research and development; and (3) the development of rapidly deployable, high-quality systems, goods, and services. Throughout the Cold War, this strategy, albeit not always perfectly implemented, proved effective because the United States had both the commitment and the resources to maintain the superior technological infrastructures and capabilities needed, and because the compelling national security mission and technical challenges attracted top STEM talent. Many new technologies were created to serve national security purposes. Remarkably, this overarching strategy did not change for nearly half a century, the longest enduring strategy in U.S. history.

However, in the 1990s a stream of global changes disrupted this strategy of complete technological superiority. Though these changes derived from different sources, they were often interrelated and carried by the irrepressible current of globalization. A major change in national and regional relationships and alliances followed the collapse of the USSR and the Warsaw Pact and the substantial expansion in the number of contributors to and customers in the global economy. Relationships between countries could be collaborative or adversarial depending on the particular issue. The Internet became the primary and inexpensive means of communication and commerce, and search engines such as Google made information freely accessible to essentially everyone worldwide, a departure from the goal of information control during the Cold War. The globalization of talent, business, and markets became the norm whereby even the smallest businesses could become global players. The rise and strength of emerging economies became significant attractors of businesses, markets, and growth in a tightly connected, interdependent global economy. China became the world's second largest economy in 2010, 3 years after a prediction published in *Rising Above the Gathering Storm* that it would occur 10 years hence in 2016 (NAS, NAE, IOM, 2007, Figure 9.1 and p. 206). Accelerating change shortened the life cycle of goods, services, and knowledge and pressed industry to move products to the marketplace more quickly, placing a premium on having a workforce prepared with needed capabilities. Accelerating change required the military to respond more quickly, more often, and in new ways to combat new and often unknown, non-state adversaries.

Since the 1990s, scientific and technological developments for national security are increasingly *not* located in the United States (National Research Council, 2009; NRAC, 2010). The United States and DOD do not control all of the technology used for military purposes. In fact, this technology is increasingly originating in commercial endeavors. The news media remind us almost daily that information, even ostensibly secure information, can no longer be controlled reliably.

The United States does not lead in all areas of science and technology, and it may not be possible to regain that leadership. The impact factor of research publications has long been held up as an indicator of a nation's leadership in science and technology. After ranking first globally in research publication impact for decades, the United States slipped to third in 2011, following the United Kingdom and Germany (Figure 1-1) despite maintaining the highest national investment in research (Marshall and Travis, 2011). The 2010-2011 World Economic Forum in Davos ranked the U.S. economic competitiveness fourth among 139 countries after it had ranked second a year earlier and first a year before that (World Economic Forum, 2010, pp. 21 and 421). The Information Technology and Innovation Foundation ranked the United States sixth in global innovation and competitiveness in 2009, down from first in 1999 and earlier (Atkinson and Andes, 2009).

In 2008 the percentage of engineering graduates among all university graduates in the United States remained among the lowest in the world, at 4.4 percent. The percentages of engineering graduates in some other countries are as follows: Germany (12 percent), U.K. (6 percent), Finland (15 percent), France (14 percent), China (31 percent), Japan (17 percent), S. Korea (25 percent), Taiwan (24 percent), Israel (10 percent), Russia (10 percent), and Singapore (34 percent). The global average percentage of engineering graduates among the 93 countries shown in an analysis by the National Science Foundation (NSF) (National Science Board, 2012, Appendix Table 2-32) is 13 percent, three times the U.S. rate. Among all 93 countries in the referenced NSF data, Mozambique most closely resembles the United States, with engineering graduates at 4.5 percent and science and engineering graduates at 32 percent. Only 14 countries in the NSF analysis graduate a lower percentage of engineers than the United States: Bangladesh, Brunei, Burundi, Cambodia, Cameroon, Cuba, Gambia, Guyana, Lesotho, Luxembourg, Madagascar, Namibia, Saudi Arabia, and Swaziland.

Since WWII, attracting the very top students from abroad to enroll in U.S. graduate programs and then stay on in the United States to develop their engineering careers has largely compensated for the shortfall in U.S.-born

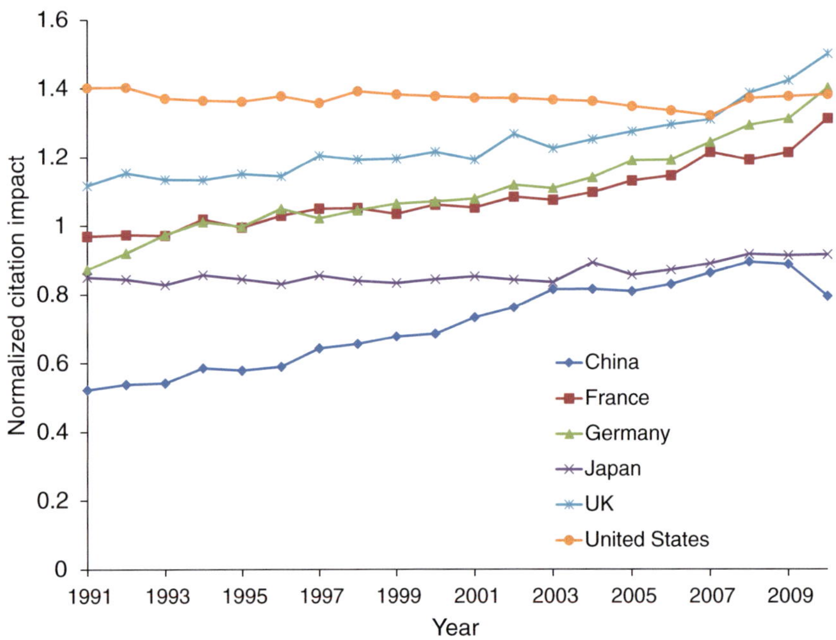

FIGURE 1-1 Global research publication impact.
NOTE: Counts are national averages and are normalized to the average number of citations in the respective research discipline.
SOURCE: Marshall and Travis (2011).

engineering talent available to the workforce. The United States was able to attract the most qualified international talent by being the most technologically advanced country, by having a growing economy, by possessing a disproportionate share of the world's finest research universities, and by committing to a world-leading higher education and research culture with strong financial support by the U.S. government (e.g., through research assistantships, funding for basic research, and support for research equipment). With less than 5 percent of the global population but a quarter of its economy, the United States had the rare opportunity to attract the very best of the global science and engineering talent pool to its workforce, and it capitalized on this remarkable, though unsustainable, circumstance. In 2006 the most likely undergraduate alma mater of a U.S. PhD graduate in science and engineering was Tsinghua University in Beijing, followed closely by Peking University (Mervis, 2008). The University of California, Berkeley, ranked third after having held first place for all earlier rankings. Ranked a close fourth, and rising rapidly, was Seoul National University in Korea. In 2010, the most recent year for which data were available, Berkeley had regained the top spot, principally because students from Tsinghua and Beijing Universities, graduating the top students in China, are not enrolling in U.S. PhD programs as they did earlier (Figure 1-2). The 2010-2011 World Economic Forum ranked the U.S. undergraduate higher education system 26th out of 139 countries and secondary education in mathematics and science 52nd (World Economic Forum, 2010, pp. 21 and 421).

The United States is no longer the beneficiary of uncompetitive higher education and job opportunities abroad that had earlier inspired large numbers of international students and scholars to come to and remain in America. As the standards of higher education and job opportunities abroad continue to rise, the competition in recruiting top talent to the United States can only increase. The emerging economies of China and India now offer attractive opportunities for wealth and professional growth for scientists and engineers. International universities and businesses are recruiting international students (and faculty) with first-class research facilities and opportunities, a force with which the United States has never had to compete. And while the numbers of students from India and China coming to the United States for graduate study remain high (Figure 1-3) and while they often pay their own way, a look below the surface shows that those attending U.S. universities are no longer at the very top of their national talent pool as they once were.[1] Attractive opportunities in other countries have made recruitment of the top talent a competitive challenge that the United States did not face in the past.

An April 2011 report from the Kauffman Foundation (Wadhwa et al., 2011) points to indicators that Indian and Chinese residents in the United States are returning home in increasing numbers because of economic opportunities, access to local markets, and family ties. The Chinese Ministry of Education estimated that the number of overseas returnees to China in 2009 increased 56 percent over the previous year, and in 2010 the number increased another 33 percent over 2009 to a global total of 134,800 (*China Daily*, 2010, 2011). Over 80 percent of Chinese returnees and 70 percent of Indian returnees indicated that the opportunity to start a business was more favorable at home than in the United States. Many other countries, such as Taiwan, Singapore, and Ireland, are recruiting high-quality S&T talents from abroad.

The challenges for the United States in the 21st century environment outlined above are significant, though until recently the U.S. public and government tended to look inward and did not show evidence of comprehending the seriousness of such challenges.

TWENTY-FIRST CENTURY DOD STEM WORKFORCE ENVIRONMENT

In this rapidly changing world, the technologies of importance to the military are created globally in increasing numbers, including those widely employed in U.S. weapons systems. The development of a weapons system—including all components, tools, and raw materials—entirely in the United States is uncommon if not altogether nonexistent. Efforts to predict the technologies that will be most needed by the military beyond the near term have always been unreliable. Resource limitations and the expanding range of S&T developments globally will nonetheless require DOD to select the S&T areas where it will maintain technological superiority. However, it will also be important for DOD to retain the capacity to ramp up programs quickly to become competitive in

[1]For example, the number of graduates from India's premier technical university, the Indian Institutes of Technology, who seek graduate study and research opportunities in the United States declined from 80 percent in 1997 to just 16 percent in 2011. See the *Times of India* (2011).

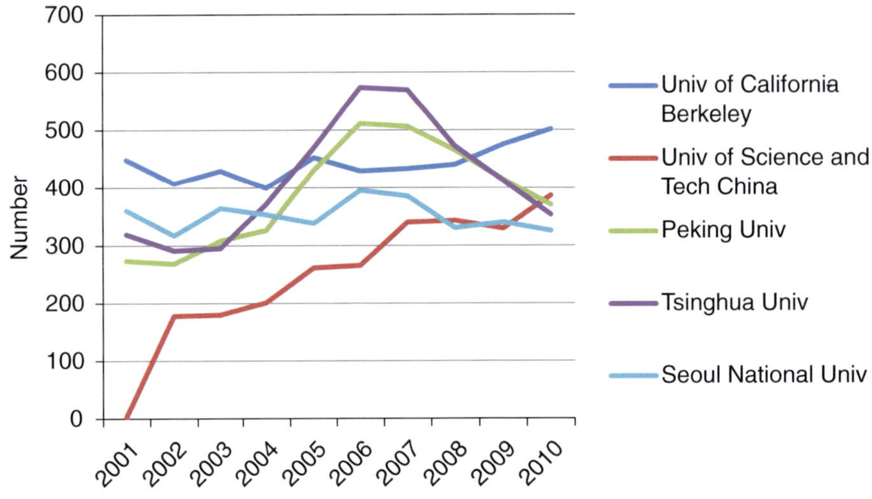

FIGURE 1-2 Baccalaureate origins of PhDs from the largest feeder schools, 2001-2010.
SOURCE: National Center for Science and Engineering Statistics, National Science Foundation.

emerging areas by making targeted R&D investments to maintain core competencies and to be highly adaptable in its management practices.

The environment for the DOD STEM workforce, including its military and civilian employees and its private sector contractors, has changed radically since 1991 and the end of the Cold War. During the nearly half-century of the Cold War, the DOD STEM workforce took on the clear and compelling national security mission to maintain technological superiority in weapons and military systems. National security was widely accepted and supported as the highest priority for the United States. No other national issue has galvanized public support over such an extended period. The national security mission attracted a career-committed workforce with the highest technical

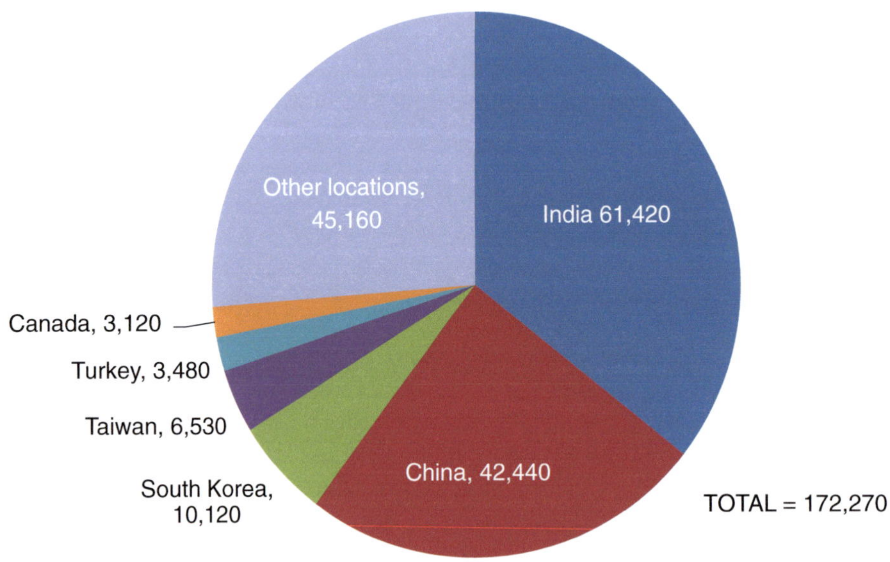

FIGURE 1-3 Foreign graduate students enrolled in S&E fields, 2009.
SOURCE: National Science Board (2012), Appendix Table 2-24.

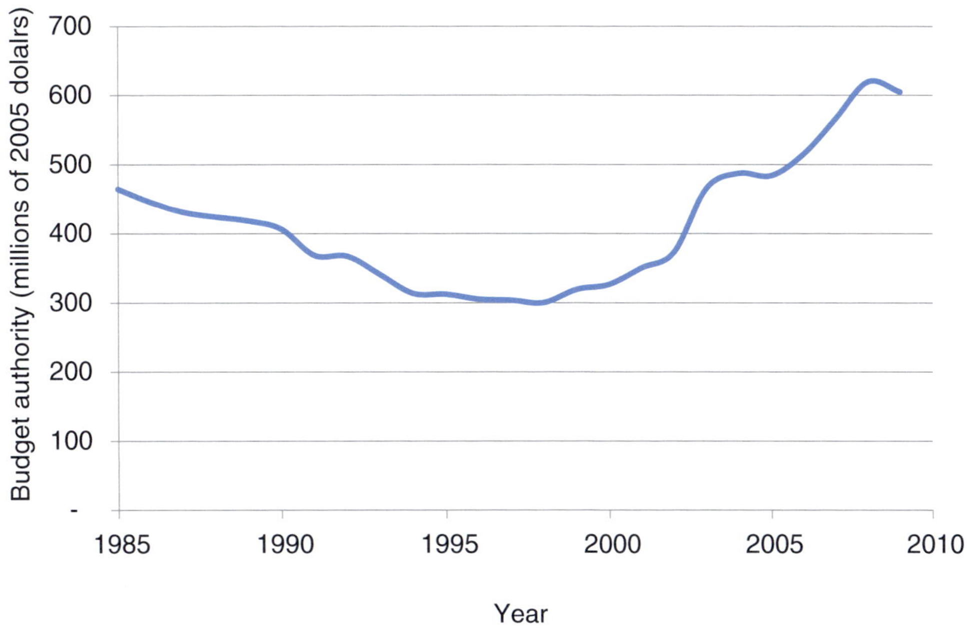

FIGURE 1-4 Total budget authority of DOD military programs, 1985-2009 (in constant 2005 dollars).
NOTE: Includes base budget and overseas contingency operations.
SOURCE: OMB historical tables. Available at http://www.whitehouse.gov/sites/default/files/omb/budget/fy2013/assets/budauth.xls.

capabilities and devotion to the security challenge. Because the newest technologies often served national security needs, the technical work itself attracted STEM employees of the highest technical capabilities.

The culture of the DOD STEM workforce during the Cold War was set by the widely understood, long-standing foundation of *continuous national support*, *workforce stability*, *workforce quality*, *technical challenge,* and *national service*. Those recruited to the workforce knew what to expect and what was expected of them. That stable foundation was disrupted by the stream of global changes noted above following the Cold War. The United States shifted national priorities toward domestic and social issues rather than foreign policy, and within foreign policy toward economic rather than political and military issues (Auger, 1997). Some of the concerns that received increasing attention included the demands of expanding populations for social services, the decline of the industrial base, the retraining of the workforce, the rebuilding of cities, the provision of clean, affordable energy, the protection of the environment, needed attention to addressing race, gender, and class inequalities, and the ability to compete in international markets (Crotty, 1995). Military spending declined substantially between 1985 and 1993, remained relatively flat until 1999, and then increased dramatically following the attacks on New York and Washington on September 11, 2001 (Figure 1-4). The reductions in DOD workforce and programs in the early 1990s signaled a transition to a new, as yet undefined culture for DOD S&T and its workforce, with a recent study finding that in the Air Force "career fields requiring a STEM degree may have experienced below-average retention or promotion rates" (National Research Council, 2010). The recession of 2008, the ongoing troop withdrawals from the Middle East, and the current national debt crisis will result in substantial DOD budget and program reductions, thereby adding uncertainty to the new culture for DOD S&T and the DOD STEM workforce.

The greatest emerging threat to U.S. national security today is not as universally apparent and as compelling as the possibility of thermonuclear war was during the Cold War. The possible future adversaries, their geographical region, and the type and the scale of conflicts are also less certain. The *stability* of the adversary, the technical challenges, and the compelling mission that characterized the national security culture throughout the Cold War do not characterize today's environment. *Adaptability* has replaced *stability* for today's challenges in workforce preparation and technical focus.

THE CURRENT STUDY

This study by the National Academy of Engineering (NAE) and the National Research Council (NRC) was requested by the Honorable Zachary J. Lemnios, Assistant Secretary of Defense for Research and Engineering. Over an 18-month period, the NRC's Committee on STEM Workforce Needs for the U.S. Department of Defense and the U.S. Defense Industrial Base (Appendix A) convened four meetings dedicated in part to open, information-gathering sessions and two closed meetings dedicated to deliberation and writing. Among the former was a workshop held on August 1 and 2, 2011, in Rosslyn, Virginia, to gather a broad range of views from the public and private sectors, including major defense contractors and nongovernmental organizations (NGOs), all of whom are stakeholders in the future STEM workforce. A report issued in November 2011 summarized the views expressed by individual workshop participants. An interim report was issued in June 2012 for the purpose of assisting the ASD(R&E) with its fiscal year (FY) 2014 planning process and with laying the groundwork for future years (National Research Council, 2012). Overall, this 18-month study has assessed the STEM capabilities that DOD needs in order to meet its goals, objectives, and priorities; to assess whether the current DOD workforce and strategy will meet those needs; and to identify and evaluate options and recommend strategies that the department could use to help meet its future STEM needs. The statement of task for the study is given in Box 1-1.

BOX 1-1
Statement of Task

A joint National Academy of Engineering (NAE)-National Research Council (NRC) study committee will assess the science, technology, engineering, and mathematics (STEM) capabilities that the U.S. Department of Defense (DOD) needs to meet its goals, objectives, and priorities; assess whether the current DOD workforce and strategy will meet those needs; and identify and evaluate options and recommend strategies that the department could use to help meet its future STEM needs.

The study work scope will involve five major tasks:

1. Review the current and projected STEM workforce demands over the next five years relevant to DOD needs and to the needs of the industrial base supporting DOD programs and missions, including an overview by science and engineering discipline, quality, and skill level.
2. Provide an assessment of current limitations to meeting these needs over the next five years and an analysis of observations by recognized experts on the forces shaping limitations on future needs.
3. Review alternative options for overcoming identified limiting factors and other impediments to fulfilling near-term DOD STEM needs.
4. Identify emerging science and technology fields that will likely have significant impact on the DOD and national needs over the next 5-15 years and where targeted national investments could have the most impact on developing human resources in the identified fields.
5. Provide an overview and analysis of expert views on the capacity of the nation's higher education enterprise in meeting the necessary scale and scope of the STEM workforce needs for DOD and the U.S. defense industrial base.

The study committee will convene a two-day public workshop on U.S. defense-related workforce needs. The workshop will feature invited expert presentations and discussions. The committee will develop the workshop agenda, select and invite speakers and discussants, and moderate the discussions. Experts to be invited to participate in the workshop will be drawn from the membership of prior NRC studies and related activities, the public and private sectors, and from academic organizations. Following the conclusion of the workshop, a summary report of the event will be prepared by the committee. There will be one administrative progress report and one interim report, as well as a final consensus report based on the committee's work on the five study tasks, including the information presented in the workshop.

The balance of this report is organized as follows: Chapter 2 discusses rapidly evolving areas of science and engineering having potential for significant impact on DOD planning and operations. Chapter 3 elucidates trends in the overall STEM labor force and discusses most likely future scenarios for DOD. Chapter 4 discusses the limitations faced by DOD and the industrial base in meeting its STEM workforce needs. Chapter 5 discusses the educational institutions that feed and maintain DOD's STEM workforce and some impediments DOD faces within this enterprise. Lastly, Chapter 6 offers a perspective on ensuring an adequate workforce capability in an uncertain future.

REFERENCES

Atkinson, R., and S. Andes. 2009. The Atlantic Century: Benchmarking EU and U.S. Innovation and Competitiveness. Washington, D.C: Information Technology and Innovation Foundation.

Auger, V.A. 1997. The National Security Council System After the Cold War In U.S. Foreign Policy after the Cold War, edited by R.B. Ripley and J.M. Lindsay. Pittsburgh, Pa.: University of Pittsburgh Press.

China Daily. 2010. "Sea turtles" swimming back in larger numbers. China Daily. Available at http://www.chinadaily.com.cn/m/eduonline/2010-03/30/content_9661084.htm (accessed October 3, 2012).

China Daily. 2011. Expat student numbers rise in China. China Daily. Available at http://www.chinadaily.com.cn/china/2011-03/04/content_12113282.htm (accessed October 3, 2012).

Crotty, W. 1995. Post-Cold War Policy: The Social and Domestic Context. Chicago: Nelson Hall.

Marshall, E., and J. Travis. 2011. U.K. scientific papers rank first in citations. Science 334:443.

Mervis, J. 2008. U.S. graduate training—Top Ph.D. feeder schools are now Chinese. Science 321(5886):185-185.

National Academy of Sciences, National Academy of Engineering, and Institute of Medicine. 2007. Rising Above the Gathering Storm: Energizing and Employing America for a Brighter Economic Future. Washington, D.C.: The National Academies Press.

National Research Council. 2009. Beyond "Fortress America": National Security Controls on Science and Technology in a Globalized World. Washington, D.C.: The National Academies Press.

National Research Council. 2010. Examination of the U.S. Air Force's Science, Technology, Engineering, and Mathematics (STEM) Workforce Needs in the Future and Its Strategy to Meet Those Needs. Washington, D.C.: The National Academies Press.

National Research Council. 2012. An Interim Report on Assuring DOD a Strong Science, Technology, Engineering, and Mathematics (STEM) Workforce. Washington, D.C.: The National Academies Press.

National Science Board. 2012. Science and Engineering Indicators 2012. Arlington Va.: National Science Foundation.

NRAC. 2010. Status and Future of the Naval R&D Establishment. National Research Advisory Committee.

Times of India. 2011. Drastic drop in number of IIT graduates going to US. Times of India, March 29.

Wadhwa, V., S. Jain, A. Saxenian, G. Gereffi, and H. Wang. 2011. The Grass Is Indeed Greener in India and China for Returnee Entrepreneurs. Kansas City, Mo.: Ewing Marion Kauffman Foundation.

World Economic Forum. 2010. The Global Competitiveness Report 2010-2011. Geneva: World Economic Forum.

2

Emerging Science and Technology Fields

INTRODUCTION

Over the next 5-15 years, science, technology, engineering, and mathematics (STEM) resources—people and funding—will be needed by the Department of Defense (DOD) to fill at least four critical needs: (1) creating new capabilities, (2) identifying threats from new capabilities created by potential adversaries, (3) evaluating and advising decision makers on technology to best increase military readiness and functionality, and (4) providing the "intelligent customer," a knowledgeable interface with industrial partners, to obtain the best technology at reasonable cost. The STEM skills required for these functions include the spectrum of capabilities from basic research to advanced engineering.

The sections below, which are not meant to be exhaustive, identify rapidly evolving areas of science and engineering with a potential for high impact on future DOD operations. STEM personnel will create, recognize, and exploit breakthrough discoveries, engineer prototypes and operational versions for military use, and integrate them into systems controlled by humans. Although it dos not minimize the requirement for continued advancement in traditional militarily important areas such as corrosion and structural fatigue, this chapter focuses on providing examples of a few rapidly expanding areas that can provide far-reaching and pervasive new technologies that will have implications for the STEM skills needed by DOD and the industrial base (Harris, 2011). The committee discusses five cutting-edge science and engineering technological systems that are likely to impact DOD capability, including (1) information technology, (2) autonomous systems, (3) systems biology, (4) innovative materials, and (5) efficient manufacturing. These interdisciplinary technologies require basic research expertise interwoven with engineering innovation to realize the potential for new DOD capabilities. For example, the section on innovative materials discusses applications of nanotechnology; the section on systems biology treats the human-machine interface; and the sections on autonomous systems and systems biology discuss energy.

INFORMATION TECHNOLOGY

Information technology is pervasive in DOD systems and has been the enabler of many military capabilities. The potential for increased innovation in capabilities is substantial. A few key areas highlight potential opportunities for advancing critical capabilities: data mining, cybersecurity, cloud computing, and communications technology.

Data Mining

The proliferation of data from sensors and intelligence gathering is overwhelming to humans. The computing activity known as *data mining* uses statistical and artificial intelligence techniques to extract useful information from databases of ever expanding size, where manual interpretation of data is impossible. The data mining task includes automatic or semiautomatic analysis of data for extraction of information found in operationally relevant patterns. Individuals engaged in data mining require knowledge of computer science, large database management, statistics, and relevant subject matter expertise. For instance, to extract useful associations out of telephone chatter from a foreign battlefield will require knowledge of language and local customs. Data mining has been extensively used in civilian environments, including market analysis, customer behavior, human genetics, spatial analysis of geophysical data, and even in high-energy physics experiments. While the field is expanding very rapidly, each use of machine learning must be grounded in deep understanding of the subject domain.

Network science, in particular dynamic link analysis, is a rapidly developing area related to data mining that is emerging as a distinct, multidisciplinary field. The combinatoric complexity of networks has led to alternative statistical approaches that go beyond static analysis. The Internet and more specialized communications systems are highly dynamic. Understanding the effects of those dynamics will be key to addresing significant problem areas such as needle-in-haystack issues, detection of anomalous behavior, and defending against cyber threats (and developing offensive cyber capabilities).

Cybersecurity

As the military, and society generally, have become dependent on information systems, communications, and computing, cybersecurity has become a critical capability. Even the cyber vulnerabilities of some civil infrastructure threaten assured operations outside military theaters. Military concerns about cybersecurity are not limited to military-owned infrastructure. It is in the interest of the military that the civilian STEM workforce be knowledgeable about the best information assurance techniques.

Cybersecurity research challenges include ensuring the integrity of data, controlling access to sensitive information, making data accessible when needed, protecting privacy, preventing intrusion, preventing access to data that is unencrypted while it is being processed, and managing degraded information systems to effectively serve priority mission needs. In addition, it is a challenge to know whether combining multiple sources of data increases the sensitivity of the merged data, when, for example, personal identity associated with a record might be inferred.

The cybersecurity STEM workforce will need to apply new approaches in algorithms, hardware and software architectures, and the design and engineering of complex, secure systems. This is particularly complicated by the fact that education and training programs outside the intelligence and military communities address only defensive cybersecurity. It is incumbent on the intelligence community to continue to explore ways to partner with industry and with educational institutions to provide the STEM workforce a strong background in effective approaches to cybersecurity.

Cloud Computing

A recent development in computing technology is the centralization of storage and heavy-duty computing capabilities in locations separate from the user's PC. In many ways, this development is reminiscent of the early days of computing, when a user's desk had only a terminal and all the storage and computing were executed on a mainframe computer located somewhere else in the building. The difference between the old and the new is the communication protocols and bandwidth that are available. Cloud computing, as opposed to using a large central mainframe, relies on sharing common hardware resources such as memory and CPU that are accessed via the Internet.

The driver for cloud computing is the need to get users' applications loaded and running faster at considerably lower cost, reduced local maintenance, and higher reliability of resources including servers, storage, and networks. With the availability of handheld devices such as smart phones and notepad computers, cloud computing is a growing part of the IT infrastructure. Cloud computing provides a cost-effective alternative to the existing paradigm of relying on local computing capability.

Both the intelligence community and the military are rapidly adopting cloud infrastructures because of their efficiency, flexibility (especially when scaling compute activities), and more centralized administration. Clouds provide a centralized information infrastructure that offers distributed access, making assured cyber protection even more vital.

Security issues include controlling access to sensitive data, segregating data, insuring privacy and data integrity (including during data processing), and preventing intrusion. The inherent efficiency and flexibility afforded by cloud computing have already resulted in its rapid acceptance in the commercial arena, and the DOD is exploring potential applications.[1]

Communications Technology

Current and future military communications systems rely heavily on mobile communications systems. These systems can integrate individuals, autonomous units, and command nodes. Essential elements are throughput capacity and security. New advances in optical communications are also creating faster logic elements and broad bandwidth communications with reduced power for both fixed and mobile systems. One essential characteristic for national security is that the networks should be fail-safe or fail-soft. To meet this requirement, the networks reconfigure or reassemble autonomously to compensate for a failure in part of the system.

Application to Training

One application of IT is training warfighters in new skills and doctrine. Uncertainty about the types of military engagements the United States is likely to face in the next decade creates an urgent requirement for "anywhere, anytime" training. The readiness of U.S. military and diplomatic establishments to engage in situations that range from major confrontations in the Pacific, to terrorist attacks on the United States or our allies by non-state groups, to missile attacks or dirty bomb assaults on U.S. population centers, requires continuous training of combatant commands and continental United States forces. With the rapid development of worldwide satellite and cellular communications and networks, the infrastructure exists to integrate these assets into a true "anywhere, anytime" training capability.

AUTONOMOUS SYSTEMS

The appearance and the acceptance of robots on the battlefield and unmanned aerial systems (UASs) in the airspace have engendered new tactical capabilities during the current Middle East conflicts (*Economist*, 2011). Steady improvements in computing, sensing, networking, and system-integration technologies have offered new capabilities for leveraging human functions with machine functions.

The emergence of autonomous systems as a key component of U.S. military power is another catalyst for integration of technical disciplines, including computing, sensing, communications, materials, and mechanical engineering. To date, however, few fully autonomous systems have become "field ready"; most deployed "autonomous systems" are actually semiautonomous, requiring an operator in the loop.

Most autonomous systems will rely on an interoperable network of manned and unmanned platforms, command and control assets, data analysis, and support functions. However, the current logistical burden associated with deployment must be significantly reduced. The trend toward smaller autonomous systems is in part driven by the potential advantage of their reduced support demands for operation and maintenance.

Autonomous systems benefit from advances in conventional air, sea, ground, and space platforms and related technologies, including propulsion and advanced materials. DOD should maintain continued focus and investment

[1]The Defense Advanced Research Projects Agency issued a solicitation in June 2011 on mission-oriented resilient clouds.

in these areas, because they are critical to supporting its missions. This section does not cover advances in those traditional disciplines. Instead, new challenges and opportunities in four areas are presented to advance autonomous systems: (1) modeling and simulation, (2) operational applications, (3) sensor integration, and (4) energy and power.

Modeling and Simulation

Modeling and simulation are critical enablers for the design and operation of complex autonomous systems. Multidisciplinary optimization of vehicle, sensor, and network performance in the system-design phase relies on high-fidelity models of the system and its components. Optimization using simulations provides the means to maximize capability and flexibility relative to cost and other constraints; operational employment provides opportunity for innovative applications in a battlefield environment.

Operating autonomous systems on the battlefield in cooperation with manned assets and unpredictable enemy forces would benefit from a clear understanding of the adversary's behavior. Autonomous systems today typically operate with limited human intervention during operation. Programmed decision-making processes in the autonomous system are constrained in principle, but sensor malfunctions or disruptions to communications links can result in unpredicted responses. Mission-level modeling and simulation of the entire battle space is required to operate safely and consistently for combined manned and unmanned forces.

Operational Applications

The range of capabilities of remotely controlled robots and UASs has expanded rapidly in the last few years. These technological advances were often made in response to joint urgent operational needs[2] identified by the field commanders. The Navy is developing airborne, surface, and undersea autonomous systems for sensing and surveillance, while the Army is developing land-based robots for explosives countermeasures and materiel transport. A major factor accompanying this progress has been the recognition that many battlefield assignments can be performed more effectively, safely, and often at less cost using robots and UASs.

To date, autonomous systems such as Global Hawk and Predator have demonstrated their effectiveness on today's battlefield. As future doctrine evolves, autonomous systems will provide innovative capabilities for both the military and homeland security. The next generation of micro UASs, reconnaissance and attack UASs, soldier augmentation robots, and unmanned logistics vehicles will employ sophisticated multispectral sensors, self-organizing networks, swarming technologies, and many other bandwidth-intensive and computationally intensive technologies.

With these new capabilities, the military will likely use unmanned systems more widely in future combat as they demonstrate increasing combat effectiveness. Unmanned platforms may accompany strategic, penetrating, manned attack aircraft. Even air-to-air combat between unmanned fighters is not far beyond the reach of currently available technology. In conjunction with the growing technical capabilities of autonomous systems, DOD will develop new doctrines and concepts of operations as these systems become fully integrated into future missions.

Sensor Integration

One of the main missions of autonomous vehicles currently deployed is to carry sensors into access-denied or hard-to-reach places. To date, the primary application of sensor suites has been for intelligence, surveillance, and reconnaissance (ISR). Remotely piloted aircraft can patrol a designated airspace continuously for 24 hours or more, taking imagery and other data to identify threats and protect forces. Such ISR support will continue to be a major mission in future conflicts; this particular application will be facilitated by improved data mining technology.

[2]DOD maintains a Joint Urgent Operational Needs Fund (JUONF), which "provides resources for urgent and compelling requirements that will prevent critical mission failure or casualties." See http://comptroller.defense.gov/defbudget/fy2012/budget_justification/pdfs/02_Procurement/JUONF_PB12_PDW_Final.pdf.

Integrating sensors on autonomous systems to support ISR tasks requires a precise understanding of the vehicle's capabilities and operational needs. Equipping a remotely piloted vehicle (RPV) or unmanned ground vehicle (UGV) with sensor suites that might include cameras or spectrometers, antennas for signal collection and data transmission, GPS, and radar requires building many sensitive systems into a vehicle while ensuring that all of them receive adequate power and cooling and do not interfere with each other. The challenge becomes even greater with emerging micro-RPVs and small ground-based robots. Early examples are already being tested for operations, including ISR, in urban areas.

The skills required to address these ISR, power, and structure challenges fall within existing disciplines. Autonomous systems, however, pack an unusual number of disparate, highly integrated systems into each vehicle. Fully autonomous operation of these vehicles also includes integrating machine logic and smart sensors with an understanding of the physics of the vehicle and its surroundings in order to "see" and navigate through the world. Providing this limited "cognitive" ability to next-generation autonomous systems while continuing to collecting operational sensor data frees system operators from mundane supervision tasks and allows the warfighter to focus on the mission, not on operating the vehicle.

Energy and Power

One key advantage of autonomous systems is their long-duration operation, far beyond human endurance. Remotely piloted aircraft can provide a steady watch for a day or more, while small, unmanned ocean "gliders" have operated for weeks at a time. The unblinking pictures of the battlefield provided by these vehicles can offer insights into the operations and tactics of adversaries. Nevertheless, the operational time may be limited by the onboard energy supply, and this problem becomes more acute with decreasing platform size.

Enhancing the duration of continuous autonomous systems requires advances in energy storage and power generation, from higher-capacity batteries to more efficient combustion engines. Commercial industry is pushing for many similar advances. Demand for energy and power solutions will generate innovative solutions. One such area is the move toward a modern electric transmission and distribution system or "smart grid" that is secure, is self-healing, and optimizes assets and operates efficiently (National Research Council, 2009a). The DOD will need a STEM workforce capable of understanding and assessing a range of technologies from multiple fields if it is to remain a smart buyer and integrator.

Improvements in traditional engines, turbines, and other propulsion systems will provide one avenue to increase the duration and mission utility of autonomous systems. Coupled with smart vehicle design, internal combustion engines and turbines are likely to provide power for many of tomorrow's systems. The use of standard ground and aviation fuels will help autonomous systems integrate into existing logistics chains. For smaller systems, solar cells, fuel cells (including those specified for biofuels), and advanced batteries may provide sufficient power.

Recent efforts in autonomous systems, like DARPA's ISIS (Figure 2-1) and Vulture programs, seek multi-year endurance (*Defense Industry Daily*, 2010; Eaton, 2010). Independent operation for years at a time has traditionally been the province of satellites and space probes. Creating aircraft capable of this feat requires self-contained energy systems that can generate enough power from the environment to sustain flight and maneuverability. Multi-year-endurance autonomous systems must harvest and store energy from the environment with systems designed to operate without maintenance or downtime.

Achieving long-duration operation in autonomous systems may also rely on unique propulsion mechanisms and operational concepts. Ocean gliders, for example, ride on currents and vary their buoyancy to achieve controlled forward motion (National Oceanic and Atmospheric Administration, 2009). Rather than simply harvesting energy from the environment to generate electricity, designers can choose to create a vehicle that relies on the unique features of its operating environment for high-efficiency propulsion. Both aerial gliders and airships utilize air currents for propulsion. Hybrid autonomous vehicles that combine mechanical propulsion and gliding are also under development.

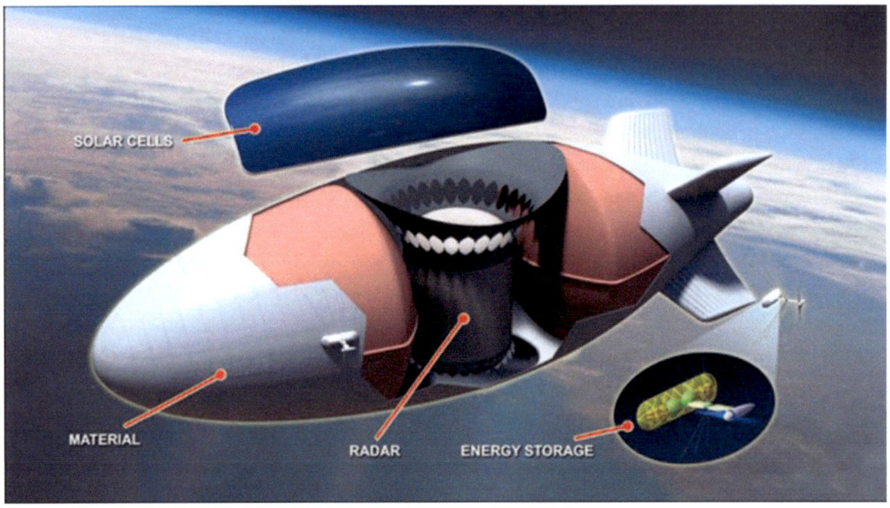

FIGURE 2-1 DARPA ISIS blimp.
SOURCE: DARPA; see http://www.darpa.mil/Our_Work/STO/Programs/Integrated_Sensor_is_Structure_(ISIS).aspx).

SYSTEMS BIOLOGY

Systems biology identifies the interactions among the components of a *biological system* that give rise to the function and behavior of the system. As a paradigm, systems biology focuses on "how system properties emerge . . . the pluralism of causes and effects . . . by observing, through quantitative measures, multiple components simultaneously and by rigorous data integration with mathematical models" (Sauer et al., 2007). The development of a computational model to explain the interactions among many components and to predict the result of changing one or more of the components is critical for predicting the functional consequences. Currently, such models exist but have limited scope and so can be both predictive and verifiable only in the short term; longer-term predictions are difficult to verify and thus engender limited trust. Practically, the field of systems biology uses data from diverse experimental sources and interdisciplinary tools and personnel for characterizing the integration of complex interactions in biological systems.

Among the important consequences of the human genome project was the realization that the function of a living organism could not be explained solely by the genes involved. Other components such as proteins and metabolites are critical in the complex pathways that determine a response at the cell or organism level. Systems biology has particular promise for delivering future technology and solving real problems because of the following underlying assumptions: (1) if we can understand how complex natural systems work, we can learn how to alter specific functions (e.g., pathogenicity, human performance, bioremediation) and (2) if we can learn to alter natural systems, we can create desired functions (biofuel production, biosensing, biomanufacturing/bioprocessing).

For evaluating the potential of systems biology to advance DOD capabilities, the discussion below is presented in three parts: (1) understanding natural systems; (2) modification of natural systems to impart particular capabilities; and (3) utilizing modified natural systems.

Understanding Natural Systems

Many of the models created to understand the molecular interplay within living organisms have been developed using bacteria because they are single-cell organisms comparatively easily manipulated in the laboratory. An important nonmedical application of systems biology using selected naturally occurring organisms is bioremediation, which uses biological agents, such as bacteria or plants, to remove or neutralize contaminants in polluted soil or water. Bioremediation, in which biological agents use contaminants as a source of food and energy, often requires enriching the soil and controlling temperature and pH. Many companies have adapted microorganisms for specific soil contaminants. Similar approaches are feasible to produce bacteria that degrade contaminants of military concern, such as explosives, chemical agents, and bio-threat agents.

The tools developed for understanding microorganisms are employed to understand cell-cell interactions and the role of such interactions in much larger multicellular organisms, such as humans. For example, understanding interactions between bacterial cells is critical if we want to regulate the formation of biofilms during marine corrosion or development of dental decay. Naturally evolved cells are used in the open environment to expedite degradation of pollutants or to produce large quantities of enzymes that function at elevated temperatures or in highly acidic conditions. Understanding mammalian cell-cell interactions is critical to increasing the host response to cancer. Increasingly complex models are being developed to explain the biochemistry of inter-species interactions, such as man-bacteria, in order to understand infection and intoxication. Such studies spark hope for new approaches to medical diagnostics based on host response and to therapeutics based on new vaccines or therapeutic blockades of the biochemical cascades triggered by infection. Such knowledge is important for DOD to protect warfighters operating in areas of endemic disease as well as those exposed to biological warfare agents.

As understanding of the interplay between biological processes at the cellular level and function at the organ and whole-animal level increase, we should also be able to improve the evaluation, treatment, and prevention of problems such as traumatic brain injury, post-traumatic stress disorder, and even fatigue. Improved understanding of cognition itself is a systems biology problem amenable to molecular-level understanding. Simulation on a multi-length scale should provide new tools for improving cognition, stress management, decision making, and learning.

Systems biology at the level of the individual human has numerous other implications. In particular, understanding the cognitive process is leading to more effective training, new technologies for the man-machine interface that will be critical for utilizing robots and autonomous systems, improved information processing and decision making, and the possibility of human performance enhancement. Human-systems engineering is an evolving field that optimizes the interface between the human and his or her environment or work processes.

Already, investigators with a systems biology understanding are moving into scales larger than a single human being. Population issues such as obesity, drug addiction, violence, and mental health have a major impact on military recruiting (Burke, 2011). Interactions within populations can be modeled in much the same way as molecules within cells and used to understand phenomena such as pandemics and human health behaviors. Eventually, it may be possible to develop better models for detection of deception and for assessing intent, though models of human behavior must advance substantially before this can be realized.

Modification of Natural Systems to Impart Particular Capabilities

To date, a major impediment to successful genetic modification of organisms has been the fear, if not the realization, of unintended consequences. Cells have complex pathways that can provide alternative mechanisms to help ensure their survival if one function is changed. Systems biology provides a roadmap that identifies a multitude of molecular consequences from a single genetic change, and the tools from a subfield known as synthetic biology allow design of modified cells for a wide range of applications ranging from detection to biomanufacturing to biofuel production.

New motifs for molecular recognition are engineered into cells, enabling recognition of targets as diverse as explosives, chemical agents, pathogens, or metals. In addition, as a consequence of target binding, cells can recognize a vanishingly small amount of the target and generate an easily measurable signal (e.g., color formation). Such cells can be used over extended periods for monitoring the environment or for self-replication to provide a continuous source of miniature sensors.

As an alternative to using intact cells as sensors, cells can be designed to secrete recognition molecules that have been modified to exhibit desirable properties such as high affinity for binding to a target, anchors for incorporation on an optoelectronic surface, storage stability, or intrinsic signal-generating properties. Not only is an understanding of the possibilities for changing the systems biology of the cell required to produce the most useful molecules, but also a systems engineering understanding of the sensor as a whole and its application is necessary for the design of useful molecules.

Utilization of Modified Natural Systems as Production or Processing Facilities

For centuries, specially selected cells have been used to make beer, bread, and cheese in large quantities. Today, genetically modified cells are used for bioproduction of hormones and other therapeutics. Cells are engineered to produce the particular molecule or function of interest with high efficiency and under the required manufacturing specifications. In addition to food and pharmaceuticals, the cellular production of plastics, oils, and specialty chemicals has already been demonstrated. Neither the cellular production machinery nor the equipment for large-scale operation is as yet cost-effective for most applications.

Bioproduction is among the most important areas for biofuel development. Biofuel technology has been following the rapid advances in basic knowledge of the life sciences. This knowledge is a key to obtaining sustainable and renewable energy sources from domestic resources.

The economics and net energy balance of the ethanol fuel cycle are well understood (National Research Council, 2009b). Multiple strategies are under development to generate ethanol from the fermentation of a broad variety of cellulosic materials. Exciting work shows that altering the genome of carbon-dioxide-fixing algae makes them more efficient in the production of hydrocarbon molecules. A major task for the production of any biofuel is engineering the scale-up of successful laboratory experiments to full-scale production. Economic assessment of resource management, chemical engineering, and life-cycle costs are required at the pilot-plant stage to make reasoned decisions about the technology path to pursue.

INNOVATIVE MATERIALS

Materials science and engineering underpins many technologies critical to DOD. Emerging innovations in materials technologies are interdisciplinary, crossing boundaries between materials science, nanotechnology, biology, chemistry, and physics. This section provides a few examples of materials for energy storage, weapons systems, lightweight structures, photonics, and electronics that have application in expanding military capabilities.

The development of energy-efficient systems and devices for transportation, sensors, and platforms has had, and will continue to have, a broad impact on DOD operational capabilities. Next-generation batteries and fuel cells, for example, will enable remote operational capabilities for longer periods at lower costs and will increase portability, while reducing reliance on petroleum-based fuels (National Research Council, 2003b).

Nanotechnology has created new possibilities for engineering energy-related materials with desirable properties and novel functionalities. Examples include nanowire-based batteries and electrochemical cells exhibiting higher energy densities and improved cycling without degradation; nanoscale materials for catalysis; next-generation thermoelectric materials taking advantage of modifications in phonon transport at the nanoscale; and photovoltaic materials in which the optical and electronic properties of materials and devices better overlap with the solar spectrum to increase efficiency (Atwater and Polman, 2010; Chan et al., 2008; Li and Somorjai, 2010).

The properties of materials at high energy densities are important for high-performance weaponry, propulsion systems, ammunition, and explosives. The design of these systems and their optimization rely on understanding fundamental aspects of materials at high temperature and high pressure and shock, at length scales from atomic to bulk, and at timescales from femtoseconds to seconds. Nanostructures play an important role in these energetic materials as well, for example, through enhancing energy release by increased activation at interfaces. Nanocomposites can similarly provide enhanced reaction rates and mixing on nanometer scales for next-generation propellants and combustion devices at high energy densities with tailored release rates. Examples include nanothermite reactions in metastable intermolecular composites and in nanoscale sol-gels (National Research Council, 2003a).

Advanced structural materials research has focused on the development of high-strength, lightweight materials with many applications to DOD systems. Ductile materials include multifunctional and self-healing materials that can respond or adapt to external conditions and repair local damage such as crack formation without intervention. Other examples of advanced materials include anti-corrosive coatings for thermal protection systems or turbine engines, nanorobotic self-healing applications (sometimes bioinspired by nature), and nanomaterials for uniforms and textiles capable of shielding soldiers from harsh environmental conditions, including those associated with chemical and biological warfare. Continued R&D is required for advanced synthesis and materials processing, first-principles simulations, and the atomic/nanoscale probing of the first steps in damage/crack/defect formation and of processes at interfaces.

Studies of the interactions between light and materials are relevant to technological applications ranging from information storage and communications technology to directed-energy weapons and advanced imaging. The optical properties of materials figure critically in the development of stealth technology, and meta-materials offer new opportunities for channeling the flow of light, e.g., for cloaking applications across the entire electromagnetic spectrum. Plasmonics similarly enables control of light propagation in materials with applications to guiding light through optoelectronic chips, nanoscale lasing on chips, and high-speed computing with light, exploiting synergies between photonics, plasmonics, and electronics. Directed-energy systems based on next-generation lasers require the development of novel materials with control of the optical, thermodynamic, and electronic properties, enabling development of ultra-low-threshold and ultra-high-intensity systems.

EFFICIENT MANUFACTURING

The United States is becoming increasingly dependent on overseas manufacturing for both civilian and military goods. Lower labor costs overseas and the economies of scale achievable through global production and sales provide a substantial cost advantage to manufacturers outside the United States. Following the movement of factories, global companies are increasingly investing in research and development facilities overseas to take advantage of the proximity to their production facilities (Figure 2-2).

For national security reasons, defense production will never move overseas fully; however, at the subsystem and component levels there is already considerable foreign content in most U.S. systems. Overseas manufacturing can introduce two critical risks: (1) hardware vulnerabilities (whether malicious or unintended) can be introduced into the production process; and (2) other countries can limit access to critical products by cutting off supplies.

Economies of scale captured by commercial firms using large overseas manufacturers are of limited value to DOD and industry because defense goods are typically manufactured in small numbers. For example, the annual output of a modest commercial truck factory exceeds the total number of Humvees (i.e., the high mobility multipurpose wheeled vehicle) in the U.S. Army inventory. These small product outputs do not support large capital investments in factories using current automation technology.

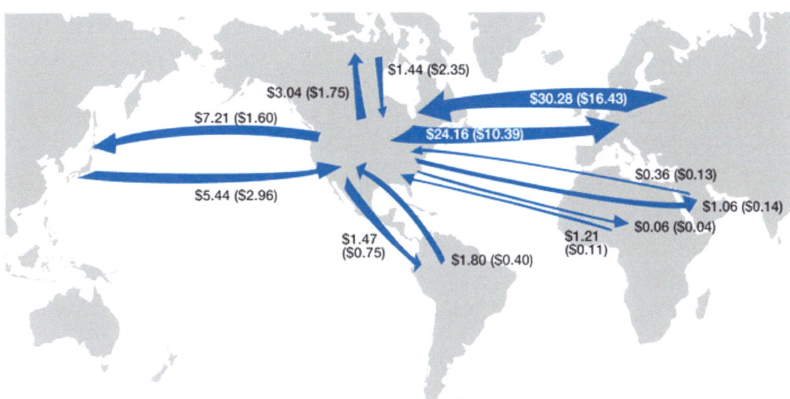

FIGURE 2-2 R&D performed in the United States by U.S. affiliates of foreign companies, by investing region, and R&D performed abroad by foreign affiliates of U.S. multinational corporations, by host region, 1998 and 2008.
NOTE: Figures in billions of current dollars. Figures in parentheses are for 1998.
SOURCE: National Science Board (2012), Figure O-6.

The DOD needs to look to other sources of innovation to improve manufacturing efficiency. Several new technologies offer that opportunity, including (1) direct manufacturing, (2) micromanufacturing, and (3) flexible robotics. A common trend that underlies advances in these three areas is the increasing use of continuous "digital threads" through design, production, and sustainment. For decades, computer-aided design has been the starting point for hardware production. Increasing automation on the production line allows the digital files generated by designers to transfer intact and electronically to manufacturing, ensuring a close agreement between as-designed and as-built products. The expansion and unification of IT systems in defense companies extends this digital thread throughout the product's life cycle. Designers, production workers, and maintainers can now use a single copy of specifications and plans, preventing discrepancies and eliminating waste associated with creating and archiving numerous copies of the same basic information.

One of the central challenges to building a STEM workforce that can continue to develop and incorporate these advances is the decline in U.S. manufacturing employment. While manufacturing advances may ultimately bring production facilities back to the United States, it is reasonable to expect that future manufacturing innovations will come from overseas universities and businesses that are closer to factories. The DOD will need to demonstrate a commitment to the U.S. industrial base and to education in manufacturing and industrial engineering to develop a workforce capable of realizing the value of these innovations.

Direct Manufacturing

Rapid direct manufacturing processes enable the production of parts from the ground up, adding new material from scratch in a step-by-step process instead of starting with a solid block and machining some of the material away. This technology, also called 3-D printing, has been used since the 1980s to produce prototype parts with accurate shapes but without durability owing to the plastic materials utilized. Based on recent advances, direct manufacturing processes can now use a dramatically expanded palette of materials. Production parts for aircraft and other complex systems, not just prototypes, are being made today with such processes. Metals and ceramics in addition to the traditional plastics offer an expaned range of material options.

Making parts directly reduces the waste associated with cutting and machining and avoids the long lead times required for cast metal parts. In addition, direct manufacturing increases the efficiency of small production facilities because it requires much less investment in tooling. For DOD systems with low production volumes, direct manufacturing offers a significant change in production efficiency and cost reduction.

Micromanufacturing

The advent of microelectromechanical systems (MEMS) in the 1980s has resulted in diverse defense applications, from accelerometers to radio-frequency (RF) electronics, cameras, and communication devices. MEMS components reduce size, weight, and power requirements. MEMS are produced using process technologies developed for the semiconductor industry. As a result, unit prices can be low despite the complexity of the systems because the cost of the production facility is shared by other, high-volume electronic components.

MEMS and similar emerging devices at the nanoscale offer a unique opportunity for DOD (Pomrenke, 1998). Realizing needed functionality in a microdevice can often be cheaper than producing it from discrete components because all of the manufacturing steps are automated. Even for relatively low-volume parts, DOD is taking advantage of fully capitalized commercial production facilities and continued advances in manufacturing efficiency and quality, driven by the needs of the global semiconductor industry.

Flexible Robotics

Traditional industrial robots are a feature of highly automated, large-scale factories around the world today. Such robots perform highly specialized tasks. At best, changing the robotic task requires substantial reprogramming and testing. At worst, changing the task requires scrapping the entire robot.

Such conventional robotic solutions are unsuitable for most defense manufacturing, in which production volumes are not large enough to justify the investment. Advances in machine vision, control systems, actuators, and the man-machine interface offer the capability to create flexible industrial robots that perform a variety of tasks. Rather than relying on parts coming down an assembly line in exactly the same place each time, an advanced, flexible robot would sense the incoming part and assess what to do with it. Flexible robots are readily reprogrammed to accommodate design or product changes, and to work alongside humans in the same way that autonomous systems operate with manned platforms on the battlefield.

Flexible robotics is on the horizon and is poised to change the design of factories. A one-time investment in advanced robots would be recouped over generations of products, bringing automated production to complex, low-volume defense hardware. DOD investments in manufacturing technology can drive this field, and an educated workforce will bring it to fruition.

STEM SKILLS RELEVANT TO THE FIVE AREAS

A requirement for expertise in information technology permeates all of the cutting-edge technology examples discussed in this chapter. Much of the innovation in information technology is occurring outside the national security community. In areas such as cloud computing and communication technologies, a particular focus on cybersecurity is necessary. The U.S. educational system is training individuals capable of creating new communications and computing strategies; however, the demand from the civilian economy is quite large, and it is not clear that DOD has the financial flexibility to compete for the highest-quality individuals in this area.

In addressing DOD skill needs for other information technology areas, both the DOD and defense contractors require teams of scientists and engineers with advanced knowledge in a range of fields plus the ability to integrate new information from those fields. For instance, applications of data mining may require individuals who are trained in computer science, data mining, linguistics, statistical analysis, cultural anthropology, and optical physics. Machine-assisted decision making is especially critical for DOD operations; critical skills required here are found in computer science, programming languages, and linguistics. Machine translation of languages requires not only expertise in software and linguistics, but also sophisticated cognizance of the current culture, and likely subculture, of those communicating in the language.

Shortages of specialists in cybersecurity have been noted by other analyses, which have estimated that thousands more offensive cyber warfare professionals may be needed, starting from a base level today of roughly 1,000 nationwide (Center for Strategic and International Studies, 2010). Further, US CYBERCOM notes it would take 18 months to train any new hires to the required level of competency. Citing concerns about offensive cyber capability in particular, Congress has recommended that DOD reorganize its current network structure to free up professionals who are otherwise serving as administrators of the numerous networks and 15,000 subnetworks (Brannen and Fryer-Biggs, 2012; Senate Committee on Armed Services, 2012).

The employment of autonomous systems by DOD will require a wide range of STEM skills. Universities today are well equipped to teach computer science, physics, mathematics, and other skills necessary for the continued development of modeling and simulation expertise. The film and video game industries are ensuring both a steady supply of students and robust competition for their talents; well-trained, talented individuals can also be attracted to work on DOD problems.

Ensuring robust communications links to control and supervise autonomous systems, providing sufficient bandwidth for desired utilization of the data they generate, and making available the computing power needed to utilize the systems' capabilities require significant advances in technology. The challenges cut across traditional disciplines, from electrical engineering and computer science to materials and optics, and even to biology for the design of control systems and models for efficient movement. The growth of bioengineering and research departments devoted to biologically inspired systems has spurred the development of computing and control systems that can manage large swarms of vehicles: the robotic equivalent of ants and bees. DOD support can encourage the growth of these and other similar initiatives to develop a robust, multidisciplinary STEM workforce to support command, control, communications, and computing (C4).

STEM education in traditional aerospace, mechanical, and electrical engineering disciplines will have to evolve to prepare students for developing multidisciplinary systems. The skills needed to address the critical problem of controlling autonomous vehicles lie at the intersection of physics, biology, information, engineering, systems, and human factors. Vehicle navigation competitions like DARPA's Grand Challenge have inspired the formation of multi-disciplinary teams to tackle these challenges (Markoff, 2007). Making autonomous vehicles energy efficient requires a multidisciplinary STEM workforce that can integrate oceanography, atmospheric science, biology, and other fields into vehicle design.

The STEM disciplines that would be central to systems biology applications such as biofuels production and bioremediation are chemical, mechanical and bioengineering; chemistry; and the biological sciences. Most of the effort in these areas will be cross-disciplinary. There does not appear to be an urgent requirement in DOD to address site remediation, but many government installations require long-term remediation of their environmentally impacted facilities. The broader application of approaches pioneered in systems biology to model and predict the responses of natural systems, including human cultures, progression of epidemics, and impact of infrastructure changes, requires teaming of information technologists, economists, and social scientists along with life scientists and engineers.

Manufacturing approaches that meet DOD requirements also require cross-disciplinary STEM talents. General needs can be fulfilled by traditional education in mechanical engineering, MEMS, 3-D printing, automated design, and materials. Cross-disciplinary improvements will address economic analyses (such as life-cycle cost projections), energy minimization, robotics, man-machine interfaces, and, almost certainly, systems engineering for both the potential products and the manufacturing systems.

Previous studies have stressed the importance of systems engineers with domain-specific knowledge who are capable of comprehending and managing all of a system's components and their interactions, and who are responsible for the design, manufacture, and operation of complex systems. Until the 1990s, government teams were involved in the front-end part of the total systems engineering process (i.e., the preplanning process); for example, the Air Force Systems Command included a structured organization with this function. Since that time, however, there has been an erosion of this embedded capability (National Research Council, 2008).

Other declines in organic STEM capacity have been documented at DOD. Following the so-called peace dividend of the 1990s, the size of the acquisition workforce declined in tandem with the procurement budget. However, when the latter increased sharply in the early 2000s, there was not a concomitant increase in the number of acquisition workers. One side effect of this decline has been that responsibility for systems engineering and integration has moved to industry, with at least one report calling for an increase in the quality of the DOD acquisition workforce (Defense Science Board, 2009). A further example: until 1998, the DOD budget included category 6.3B for systems advanced development that supported rapid prototyping programs (National Research Council, 2001).

FINDINGS AND RECOMMENDATIONS

STEM personnel are required for a wide variety of DOD R&D, acquisition, and operations. Advances in DOD capabilities in information technology, microelectronics, nanomaterials, systems biology, and direct manufacturing are critical to creating effective and affordable military systems. The potential for autonomous systems, micro-scale systems, efficient energy supplies, and improved human performance will demand input from a variety of STEM disciplines.

Finding 2-1. Advances in the technology areas relevant for future DOD capabilities, such as those described above, require knowledge from multiple disciplines. Most overlap with the commercial sphere, making DOD simply another competitor to attract high-tech talent. Teams of dedicated individuals with different knowledge bases should work together to apply cutting-edge science and engineering to solve DOD problems.

Recommendation 2-1. The STEM workforce needs training for cross-disciplinary teamwork. DOD should encourage interdisciplinary collaborations at all career stages in both academic and government laboratories through support of interdisciplinary projects, academic and on-the-job learning opportunities, and career rewards for interdisciplinary endeavors.

Finding 2-2. Transition of laboratory science and technology to deployment in DOD operations requires competent systems engineering as well as expertise in component engineering. The expected progression from graduate scientist or engineer to system engineer usually takes several years of increasing exposure to simulation and modeling, materials optimization, control and communications software development, and field testing. Today, universities often provide opportunities for undergraduate research, interdisciplinary problem solving, prototyping projects, and formal courses on system engineering. As a result, STEM graduates from many universities have some hands-on experience in cross-disciplinary projects and course work in system engineering. However, systems engineering at the scale required is performed entirely by DOD and its contractors. This understanding of systems engineering is particularly important for efficient military acquisition and preparedness for both DOD contracting and the industry responding to the government requirements.

Recommendation 2-2. The DOD should reassemble government teams to do preliminary system engineering—including affordability, capability, and sustainability—and program structuring so that the government focus is on relevant requirements when interacting with the defense industry. The industry teams also require system engineering and integration teams that can efficiently respond to the government's requirements.

Finding 2-3. Uncertainty as to the types of military engagements the United States is likely to face in the next decade creates an urgent requirement for "anywhere, anytime" training. With the rapid development of worldwide satellite and cellular communication networks, ISR (intelligence, surveillance, reconnaissance) capabilities, and modeling and simulation, the infrastructure exists to integrate these assets into a true "anywhere, anytime" training capability.

Recommendation 2-3. The DOD should initiate a major program to secure the necessary STEM-qualified government teams to deliver effective, worldwide training and to leverage information technology and ISR infrastructure to meet a mandate of "anywhere, anytime" training.

Finding 2-4. Innovative materials broadly underlie critical technology for the DOD and are essential for maintaining a technological edge. The most recent innovations in materials science are cross-disciplinary and range from fundamental science to use-inspired research and development. An emphasis by DOD on STEM education in materials science and related areas (e.g., nanotechnology, systems biology, energetics, photonics) can seed the development of new capabilities as well as new solutions to old problems.

Recommendation 2-4. The DOD should maintain expertise in materials science as broadly defined. This can be achieved in part by leveraging existing programs within DOD labs as well as at universities, and by increasing the interaction between the two. Making DOD careers attractive to the STEM workforce requires emphasis and placement of DOD resources in the entire pipeline from basic research and discovery science to applied research and product development.

Finding 2-5. The United States increasingly relies on information technologies to support its warfighters. The support provided by information technology improves the capability to respond effectively to the changing mix of challenges. Data collection, data translation, data mining, cybersecurity, and data manipulation for correct interpretation of increasing amounts of information require expertise not only in the understanding of physical sensors and advanced computing software and platforms but also expertise in linguistics and a deep understanding of local cultural nuances. Consistent with the most recent national security policy documents, the United States especially needs to increase its ability to operate in the Asian/Pacific theater. There is evidence, however, of a nationwide shortage of cybersecurity professionals with appropriate security clearances.

Recommendation 2-5. The DOD should pay special attention to the need for multidisciplinary STEM personnel to support the information technology infrastructure for defense. While individuals are being trained at universities in various specific disciplines, few individuals are trained with multidisciplinary capabilities. DOD should explore the possibility of developing multidisciplinary training in-house or in targeted university programs.

Finding 2-6. Areas of near-term technological focus with relevance to DOD's mission include the following: advanced robotics and autonomous systems; intelligence collection; cyber warfare (defensive and offensive); human-machine interactions on human terms; means to detect and neutralize biothreats; military applications of the biosciences (systems biology, biosensors, etc.); military applications of the information sciences; and nanotechnology (for innovative materials and other applications).

REFERENCES

Atwater, H.A., and A. Polman. 2010. Plasmonics for improved photovoltaic devices. Nature Materials 9(3):205-213.

Brannen, K., and Z. Fryer-Biggs. 2012. U.S. short on offensive cyber experts. Defense News, July 2.

Burke, D. 2011. Emerging Science and Technology in the Life Sciences. Presentation to the Workshop on STEM Workforce Needs for the U.S. DOD and the U.S. Defense Industrial Base, August 1, Rosslyn, Va.

Center for Strategic and International Studies (CSIS). 2010. A Human Capital Crisis in Cybersecurity. Washington, D.C.: CSIS.

Chan, C.K., H.L. Peng, G. Liu, K. McIlwrath, X.F. Zhang, R.A. Huggins, and Y. Cui. 2008. High-performance lithium battery anodes using silicon nanowires. Nature Nanotechnology 3(1):31-35.

Defense Industry Daily. 2010. DARPA's Vulture: What goes up, needn't come down. Defense Industry Daily. Available at http://www.defenseindustrydaily.com/DARPAs-Vulture-What-Goes-Up-Neednt-Come-Down-04852 (accessed June 6, 2012).

Defense Science Board. 2009. Creating an Effective National Security Industrial Base for the 21st Century: An Action Plan to Address the Coming Crisis. Arlington, Va.: U.S. Department of Defense.

Eaton, K. 2010. DARPA's Smart Blimp: Mysterious, Hovering Future of Battlefield Surveillance. Available at http://www.fastcompany.com/1589776/isis-airborne-radar-defense-surveillance-unmanned-airship-blimp (accessed June 6, 2012).

Economist, The. 2011. Flight of the drones. The Economist, October 8.

Harris, W. 2011. Presentation to the Workshop on STEM Needs for the U.S. Department of Defense and the U.S. Defense Industrial Base, August 2, Rosslyn, Va.

Li, Y.M., and G.A. Somorjai. 2010. Nanoscale advances in catalysis and energy applications. Nano Letters 10(7):2289-2295.

Markoff, J. 2007. Crashes and traffic jams in military test of robotic vehicles. New York Times, November 5.

National Oceanic and Atmospheric Administration. 2009. Ocean Glider Set to Attempt Atlantic Crossing. Available at http://oceanservice.noaa.gov/news/weeklynews/apr09/glider.html (accessed February 2, 2012).

National Research Council. 2001. Review of the U.S. Department of Defense Air, Space, and Supporting Information Systems Science and Technology Program. Washington, D.C.: National Academy Press.

National Research Council. 2003a. Frontiers in High-Energy-Density Physics: The X-Games of Contemporary Science. Washington, D.C.: The National Academies Press.

National Research Council. 2003b. Meeting the Energy Needs of Future Warriors. Washington, D.C.: The National Academies Press.

National Research Council. 2008. Pre-Milestone A and Early Phase Systems Engineering. Washington, D.C.: The National Academies Press.

National Research Council. 2009a. America's Energy Future: Technology and Transformation. Washington, D.C.: The National Academies Press.

National Research Council. 2009b. Liquid Transportation Fuels from Coal and Biomass. Washington, D.C.: The National Academies Press.

National Science Board. 2012. Science and Engineering Indicators. Arlington, Va.: National Science Foundation.

Pomrenke, G. 1998. Defense Advanced Research Projects Agency Ultra Electronics: Ultra Dense, Ultra Fast Computing Components. Available at http://www.wtec.org/loyola/nano/us_r_n_d/03_24.htm (accessed June 6, 2006).

Sauer, U., M. Heinemann, and N. Zamboni. 2007. Genetics--Getting closer to the whole picture. Science 316(5824):550-551.

Senate Committee on Armed Services. 2012. Report to Accompany S. 2354: National Defense Authorization Act for Fiscal Year 2013 (Report 112-173). Washington, D.C.: U.S. Government Printing Office.

3

The STEM Workforce in the Defense Industrial Base, Within DOD, and Overall

INTRODUCTION

This chapter highlights trends in the overall STEM workforce; discusses the current STEM talent base and the anticipated growth of STEM jobs; examines the current STEM workforce in the defense industrial base; and describes, to the extent that data permit, the DOD STEM workforce.

No single, official definition of STEM is used by DOD or the federal government. Recent studies of the U.S. STEM workforce by various government agencies differ along three key dimensions: (1) the occupations included in STEM (e.g., inclusion or exclusion of social scientists, among others), (2) the minimum stated education requirement (e.g., bachelor's degree and above versus no degree requirement), and (3) the data source used to generate the estimates (e.g., Census Bureau's American Community Survey versus Bureau of Labor Statistics' Occupational Employment Statistics). Table 3-1 (which, along with the other tables in this chapter, is given in the annex at the end of the chapter) illustrates the diverse approaches to estimating STEM employment used by three government agency units—the Bureau of Labor Statistics in the Department of Labor, the Economics and Statistics Administration in the Department of Commerce, and the National Center for Science and Engineering Statistics in the National Science Foundation. These estimates indicate that STEM employment in the United States ranges from as low as 4.75 million to as high as 8 million,[1] a difference of almost a factor of two.

No two estimates in Table 3-1 rely on the same occupation definition, education requirement, and data source! As expected, smaller estimates of the size of the STEM workforce (less than 5 million) are found in the two studies that include only those with at least a bachelor's degree. In terms of occupations, all of the estimates in Table 3-1 include biological, agricultural, and environmental life scientists; computer and mathematical scientists; engineers; and physical scientists.[2] Architects, social scientists, STEM managers, STEM postsecondary teachers, STEM sales

[1]Measures of the STEM workforce presented in this chapter are based on those working in occupations defined as STEM. However, there are other possible approaches. For example, a different approach uses educational credentials as the basis, such that the pool of STEM workers includes all those in the labor force with a STEM degree, regardless of occupation. Yet another definition is based on the level of STEM expertise workers report that their jobs require. See National Science Board (2012, Chapter 3) for additional discussion of these alternative measures of the STEM workforce.

[2]Note that the detailed occupations included in these occupational groups may differ by study. For example, the Bureau of Labor Statistics study includes actuaries in its computer and mathematical scientist category, whereas the Economics and Statistics Administration study excludes actuaries.

occupations,[3] and STEM technicians are included in one or more of the estimates. The largest estimate of approximately 8 million is found in the Bureau of Labor Statistics study, which includes the broadest set of occupations defined as STEM and imposes no education requirement.

Observation 3-1. Estimates of STEM employment in the United States vary across studies due to differences in the definitions, assumptions, and data sources utilized.

HISTORICAL TRENDS IN THE OVERALL STEM WORKFORCE

This discussion of historical trends in the overall STEM workforce relies on information available in a 2006 study released by the Commission on Professionals in Science and Technology (CPST) that examined the U.S. STEM workforce from 1950 to 2000 (CPST, 2006a). The benefit of using a single study is that it applies the same methodology over time. However, since the estimated size of the STEM workforce varies substantially across studies, the focus here is on identifying general trends in the historical STEM workforce.

The CPST study takes a broad view of the STEM workforce and includes the following occupational groups in its examination: life sciences; physical sciences; engineering; mathematics and information technology; social sciences; and science and engineering technicians. The study utilizes the decennial U.S. census to estimate the size of the STEM workforce in each decade from 1950 to 2000. These estimates are provided in Table 3-2 in the annex and illustrated in Figure 3-1. For comparison purposes, the first row of Table 3-2 provides information on the size of the overall U.S. workforce in each decade. Over the period 1950-2000, the U.S. workforce grew at an annual rate of 1.7 percent. In comparison, the overall STEM workforce grew at a considerably larger annual rate of 4.2 percent. By 2000, the overall STEM workforce was 7.7 times larger than it was in 1950 (the comparable figure for the overall U.S. workforce was 2.3). Moreover, in 1950 the STEM workforce accounted for 1.5 percent of the U.S. workforce; by 2000 this figure had increased to 5 percent.

The STEM occupational group with the largest growth during this 50 year period was mathematics and information technology, which grew at an annual rate of 10.1 percent. In 1950, mathematics and information technology constituted only a small percentage of the overall STEM workforce (2.9 percent); this figure increased to 47.5 percent by 2000. Another notable change is the steady decline in the share of the STEM workforce in engineering occupations (from close to 63 percent in 1950 to less than 27 percent in 2000). The dramatic changes in the mathematics and information technology workforce and the engineering workforce are not unrelated. These changes may be due in part to an increase in the demand for software developers relative to hardware engineers, most notably in the 1980s. Albert Endres, in an essay on the history of software engineering, notes that with the arrival of the personal computer the "traditional dominance of hardware over software ended" (Endres, 1996).

Moreover, CPST (2006a) suggests that the observed changes in STEM occupational employment over the period 1950-2000 may be due in part to changes in the way the census defined occupations. For example, according to another CPST study, computer-related occupations were added to the census in 1970 and were expanded in the 2000 Census by reclassifying a large number of electrical and electronics engineers as computer scientists (CPST, 2006b). The CPST (2006a) also notes that changes in the representation of technicians may be due to changes over time in the way these occupations are defined in the census. These occupational changes are illustrated in Figure 3-2, which shows the distribution of the STEM workforce by occupational group from 1950 to 2000.

Observation 3-2. STEM employment in the United States over the period 1950-2000 saw dramatic shifts in the distribution of the workforce across occupational groups, most notably the shift away from engineering occupations and into mathematics and information technology occupations; these changes were likely due to a combination of changes in the demand for software developers relative to hardware engineers and changes over time in the way occupations in the census were defined.

[3]Includes sales engineers and sales representatives for technical and scientific products.

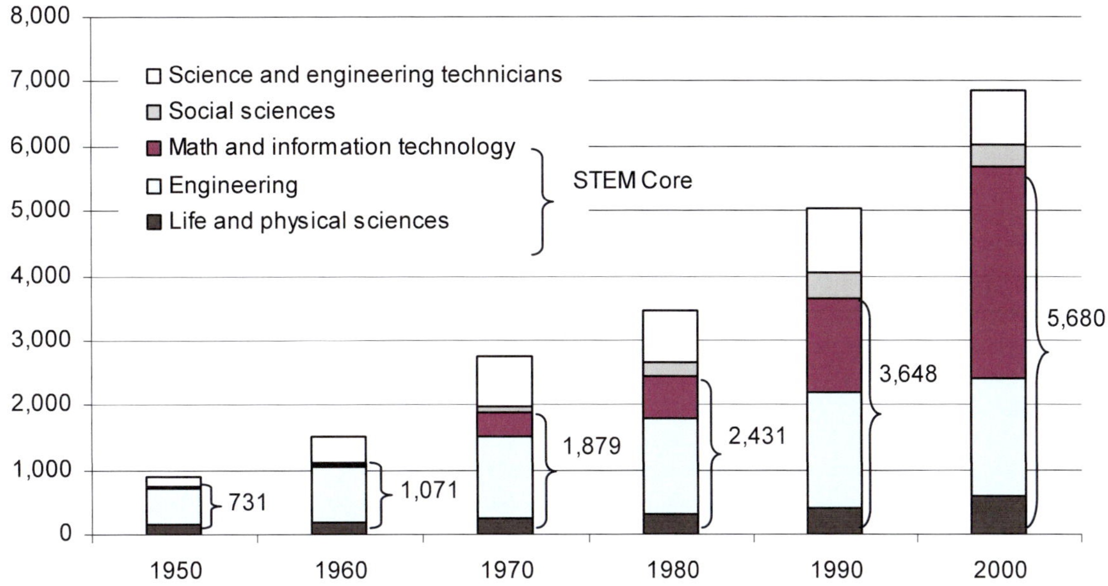

FIGURE 3-1 STEM workforce by occupational group, 1950, 1960, 1970, 1980, 1990, and 2000 (in thousands).
SOURCE: CPST (2006a, p. 3).

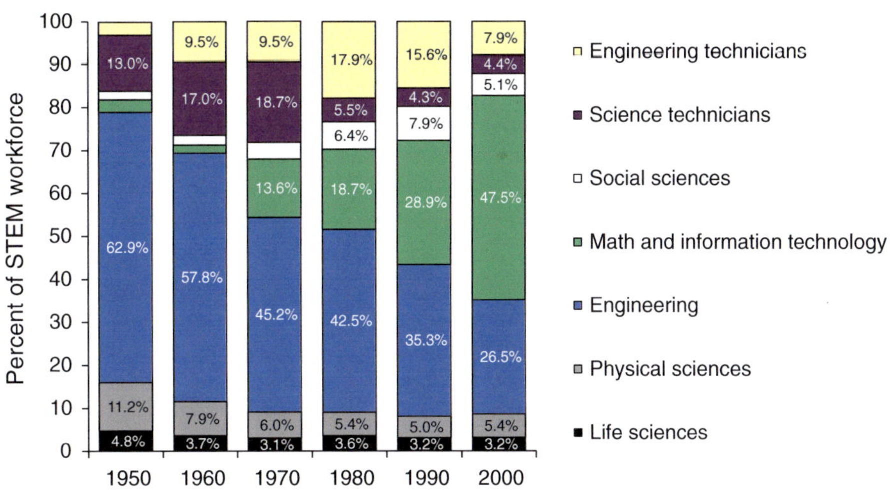

FIGURE 3-2 Distribution of STEM workforce by occupational group, 1950, 1960, 1970, 1980, 1990, and 2000.
SOURCE: Data are from CPST (2006a). Tabulations by the National Research Council.

CURRENT AND ANTICIPATED STEM LABOR MARKET

Data available from the Bureau of Labor Statistics' Employment Projections Program is used to examine the current STEM labor market in the United States and the anticipated growth of STEM jobs. This program makes ten-year employment projections every 2 years. The most recent projections estimate the number of job openings that are expected to arise between 2010 and 2020. Expected job openings in a given occupation are due to two sources: (1) job openings that are the result of expected growth in employment in an occupation (i.e., new jobs) and (2) job openings that arise from the need to replace people who are expected to leave an occupation (for example, due to retirement or career change).[4] According to the Bureau of Labor Statistics, job openings from replacement needs exceed the number of job openings from growth in roughly 80 percent of occupations (Lockard and Wolf, 2012).

It is important to note that projecting changes in employment is a difficult task. As discussed in a 2000 National Research Council workshop report on the demand and supply of doctoral scientists and engineers (National Research Council, 2000), "[a] short-term forecast, a one- or two-year forecast using annual data, can do a good job of forecasting point estimates." Long-term forecasts, such as the 10-year projections created by the Bureau of Labor Statistics, "are more complex than short-term forecasts and are more vulnerable to unanticipated changes in the economic environment" (National Research Council, 2000). In fact, the Bureau of Labor Statistics states that because "employment for many industries still had not recovered to pre-recessionary levels when the 2010-20 projections were developed," this resulted in "faster growth rates and more numerous openings than might have been expected . . . had the recession not occurred" (Bureau of Labor Statistics, 2012).

The Bureau of Labor Statistics assesses its projections periodically. In its most recent assessment of the 1996-2006 occupational projections, Wyatt (2010) asserts that the Bureau of Labor Statistics' occupational projections did better than "naïve" models (which assume that all prior trends and relationships continue). Moreover, Wyatt notes that "BLS was off by only 4.2 percent in projecting the distribution of employment among occupations"(Wyatt, 2010, p. 59). However, if one looks at the performance of the projections for specific STEM occupations, a somewhat different pattern emerges. Wyatt provides projected and actual growth between 1996 and 2006 for 18 STEM occupations. In one-third of the occupations (6 of the 18), there was either an increase in employment for the occupation when the Bureau of Labor Statistics projected a decline (2 occupations) or a decrease in employment for the occupation when the Bureau of Labor Statistics projected an increase (4 occupations). For 11 of the other 12 occupations, the ratio of the actual percentage change to the projected percentage change was at least 1.7; in the remaining case, the Bureau of Labor Statistics projected that employment would increase by 25 percent, but it increased by only 5 percent. These results suggest that the Bureau of Labor Statistics underestimated growth in many STEM occupations over the period 1996-2006. Thus, the employment projections presented here should be interpreted with caution.

One way to help mitigate concerns about the accuracy of long-term forecasts is to examine groups of occupations rather than individual occupations. According to National Research Council (2000), aggregating "typically increases forecast accuracy because random errors and movements between the occupations in the aggregate are averaged out." Using information from the Bureau of Labor Statistics' Employment Projections Program, Table 3-3 in the annex, and Figure 3-3 show 2010 estimated employment and projections to 2020 for the following STEM occupational groups—life sciences, physical sciences, engineering, mathematics and information technology, social sciences, STEM managers, and STEM technicians.[5,6] STEM managers "[p]lan, direct, or coordinate activities" (Bureau of Labor Statistics, undated) in their field of specialization and include computer and information systems managers, architectural and engineering managers, and natural science managers. Table 3-3 also includes the 2010-2020 employment growth rate, which equals the number of new jobs expected over the 10-year period as a percentage of 2010 employment; the 2010-2020 replacement rate, which equals the number of job openings due to replacement needs expected over the 10-year period as a percentage of 2010 employment; and the number of projected job openings from the combination of growth and replacement needs. For comparison purposes, the top row of Table 3-3 provides information for all occupations across the nation.

Across all occupations in the United States, employment is projected to grow from approximately 143 mil-

[4]A job opening from replacement does not include instances in which an individual moves from one company to another without changing occupation.

[5]Occupational groups are based on the following Standard Occupation Classification (SOC) codes: 19-1000 (Life Scientists); 19-2000 (Physical Scientists); 17-2000 (Engineers); 15-0000, excluding 15-2091 (Computer and Mathematical Occupations); 19-3000 (Social Scientists); 11-3021, 11-9041, 11-9121 (STEM Managers); and 15-2091, 17-3020, 17-3031, 19-4000 (STEM Technicians). Postsecondary STEM teachers are excluded since the information available from the Bureau of Labor Statistics' Employment Projections Program does not distinguish teaching field. Note that the STEM Managers category includes architectural managers since they are included in SOC code 11-9041 (Architectural and Engineering Managers) and cannot be disentangled from engineering managers.

[6]Since the definitions and data source used in this section differ from those used in CPST (2006a), care must be taken in making direct comparisons of STEM employment across the two reports.

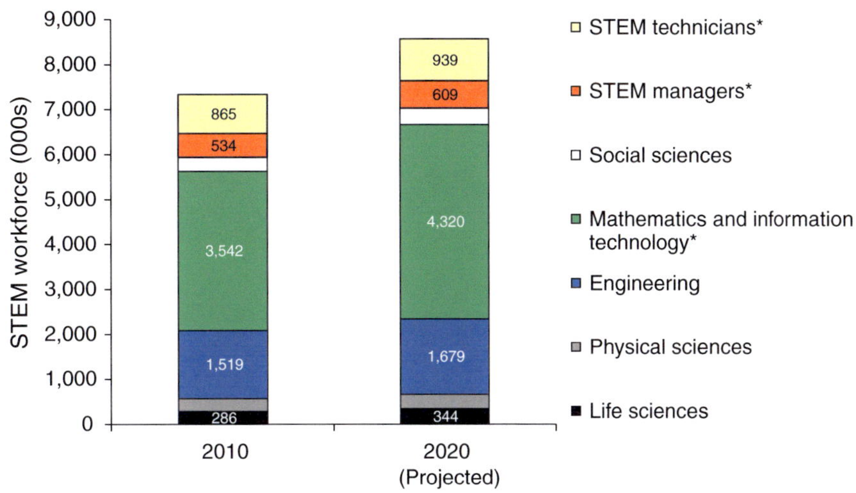

FIGURE 3-3 Employment by STEM occupational group, 2010 and 2020 (projected).
SOURCE: Data are from the Bureau of Labor Statistics, U.S. Department of Labor, Employment Projections Program (www.bls.gov/emp/). Projected job openings are due to growth and replacement needs. An asterisk (*) indicates that the information presented has been computed by the National Research Council.

lion in 2010 to more than 163 million in 2020. This increase in employment of 20 million represents a projected growth rate of 14.3 percent over this 10-year period. In addition, the estimated replacement rate over this 10-year period is 23.6 percent. Taken together, these changes are expected to result in roughly 55 million job openings during this 10-year period. Across all of the STEM occupational groups examined, employment is expected to grow by 16.9 percent over the period 2010-2020, which is slightly higher than the overall U.S. growth rate. On the other hand, the replacement rate of 21.7 percent is slightly lower than the rate for the nation. Based on these estimates, more than 2.8 million STEM job openings are expected by 2020. Roughly half of these job openings are expected to be in math and information technology occupations, with an additional 526,000 expected to be in engineering occupations.[7] It's interesting to note that the source of job openings varies by occupational group. For example, the job openings in math and information technology occupations are from a mix of employment growth and replacements, while job openings in engineering occupations are predominantly from replacements. Moreover, replacement rates are notably high in physical science occupations (30.6 percent) and social science occupations (32.1 percent).

According to these employment estimates, more than 48 percent of the current STEM workforce is in math and information technology occupations, and this figure is projected to increase to more than 50 percent by 2020. The share of the STEM workforce in engineering occupations is expected to decline to less than 20 percent by 2020. Although less dramatic, these changes follow the same general pattern of 1950-2000 STEM employment changes as presented in CPST (2006a). The projected changes in the distribution of the STEM workforce by occupational group between 2010 and 2020 are illustrated in Figure 3-4.

---

[7] Although a detailed comparison of the projected job openings in STEM occupations over the period 2010-2020 to projected degree production over this period is beyond the scope of this study, a rough calculation using degree completions data from the Integrated Postsecondary Education Data System (IPEDS) suggests that degree production will likely outpace projected job openings for all STEM occupations except mathematicians and information technology. For example, according to the projections in Table 3-3, approximately 53,000 job openings each year are expected in engineering occupations. In 2010, more than 70,000 bachelor's degrees and more than 35,000 master's degrees were awarded in the field of engineering. Moreover, there are projected to be more than 140,000 job openings each year in mathematics and information technology occupations. In 2010, however, the number of bachelor's and master's degrees awarded in math and computer sciences fields was approximately 81,000, with another 33,500 associate's degrees awarded. Note that this rough calculation does not account for diversion of STEM talent into non-STEM occupations. See Lowell et al. (2009) and Carnevale et al. (2011) for examinations of retention and diversion.

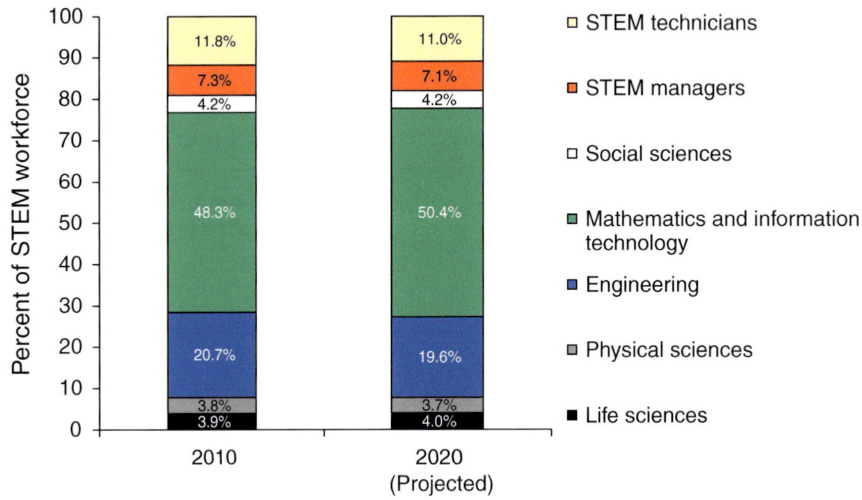

FIGURE 3-4 Distribution of STEM workforce by occupational group, 2010 and 2020 (projected).
SOURCE: Data are from the Bureau of Labor Statistics, U.S. Department of Labor, Employment Projections Program (www.bls.gov/emp/). Tabulations by the National Research Council.

Observation 3-3. More than 2.8 million STEM job openings in the United States are expected by 2020, with half of them expected to be in math and information technology occupations.

STEM WORKFORCE IN THE DEFENSE INDUSTRIAL BASE

The first step in examining the STEM workforce in the defense industrial base is to define the industries that constitute the defense industrial base.[8] According to the Department of Homeland Security, the defense industrial base includes the "private sector worldwide industrial complex with capabilities to perform research and development (R&D), design, produce, deliver, and maintain military weapon systems, subsystems, components, or parts to meet military requirements" and "includes hundreds of thousands of domestic and foreign entities and their subcontractors" (Department of Homeland Security, 2007). Due to the complexity of the defense industrial base and limitations associated with the various data sources used to estimate the size of the workforce, these estimates may, in some cases, underestimate the size of the defense industrial base STEM workforce and may, in other cases, overestimate the size.

In examining the defense industrial base, the Aerospace Industries Association (AIA) defines the aerospace and defense industry as (1) aerospace products and parts manufacturing (NAICS 3364), which consists of activities such as aircraft manufacturing, aircraft engine and engine parts manufacturing, and guided missile and space vehicle manufacturing; and (2) search, detection, and navigation instruments manufacturing (NAICS 334511), which consists of activities such as the manufacturing of aircraft instruments (except engines), flight recorders, navigational instruments and systems, radar systems and equipment, and sonar systems and equipment.[9,10] The AIA estimates aerospace and defense employment at 623,700 in 2010. A recent report prepared by Deloitte (and commissioned by the AIA) takes a broader view and also includes establishments engaged in operating a shipyard

[8]The industries discussed in this section provide services not only to the U.S. Department of Defense, but also to other government agencies (e.g., NASA, NOAA) as well as to private sector enterprises (e.g., commercial airlines).

[9]The NAICS (North American Industry Classification System) is the standard industrial classification system used by statistical agencies such as the Bureau of Labor Statistics. A comprehensive list of the most recent version of the NAICS can be found at http://www.census.gov/cgi-bin/sssd/naics/naicsrch?chart=2007.

[10]Aerospace Industries Association, Aerospace Statistics, Series 12: Total and Production Worker Employment in the Aerospace Industry (Quarterly) (http://www.aia-aerospace.org/assets/stat12.pdf).

(NAICS 336611), ordnance manufacturing (NAICS 332995), and small arms ammunition manufacturing (NAICS 332992), among others (Deloitte, 2012). The Deloitte report estimates employment at slightly more than 1 million in 2010. These two reports suggest that employment in the defense industrial base represents a relatively modest fraction of total U.S. employment (less than 1 percent).

To provide a sense of how aerospace and defense employment has changed over time, Figure 3-5 shows employment from 2005 to 2010 based on estimates provided in these two sources. Both sources show an increase in employment from 2005 until 2008, followed by a gradual decline from 2008 until 2010. The Deloitte study speculates that the recent decrease in employment "can be partially attributed to several high profile reductions in force by several major defense contractors" (Deloitte, 2012, p. 13), such as Lockheed Martin, BAE Systems, and United Space Alliance.

The AIA and Deloitte studies, however, do not specifically examine the STEM workforce in the defense industrial base. Information available from the Census Bureau's American Community Survey can be used to estimate STEM employment in the defense industrial base.[11] To generate this estimate, a broad definition of the defense industrial base (in keeping with the Deloitte study discussed earlier) is utilized. The following industries are identified as part of the defense industrial base: aircraft and parts manufacturing; aerospace products and parts manufacturing; manufacturing of navigational, measuring, electromedical, and control instruments; ship and boat building; and ordnance manufacturing.[12] Due to the way in which industry is captured in the American Community Survey, some of these industries include activities that are not specific to the defense industrial base. For example, the definition used here includes the manufacturing of medical equipment such as pacemakers and ultrasound equipment. With this caveat in mind, total employment in the defense industrial base is estimated at 1 million in 2010, which coincides with the estimate in the Deloitte study. Of this, STEM employment is about 300,000, or 30 percent of total employment.[13] These results suggest that STEM employment in the defense industrial base represents approximately 4 percent of total STEM employment.[14] Within the defense industrial base STEM workforce, the largest occupational group is engineering (60 percent), followed by mathematics and information technology (23.4 percent). Physical sciences, STEM managers, and STEM technicians each constitute 1.1 percent, 7 percent, and 8.4 percent, respectively. Life sciences and social sciences collectively represent less than 0.2 percent of the defense industrial base STEM workforce.

Information from the Bureau of Labor Statistics' Employment Projections Program can be used to examine the anticipated growth of STEM employment in the defense industrial base. However, this information is only available for a subset of the industries that constitute the defense industrial base. Specifically, information is available for aerospace products and parts manufacturing (NAICS 3364). This industry includes the manufacturing of aircraft, aircraft engines, guided missiles, space vehicles, and parts. Based on employment information provided in the AIA study, the aerospace products and parts industry accounts for approximately 77 percent of total employment in the aerospace and defense industry. This figure is a more modest 46 percent in the Deloitte study (and the estimates generated from the American Community Survey), due to the broader definition of the aerospace and defense industry used.

Table 3-4, in the annex, and Figure 3-6 present 2010 STEM employment and 2020 projected STEM employ-

[11]The American Community Survey 2010 Public Use Microdata Sample Files were used to generate these employment estimates.

[12]The following NAICS industry codes in the American Community Survey 2010 Public Use Microdata Sample Files were used to identify these industries: 33641M1, 33641M2, 3345, 3366, 33299M.

[13]The following Standard Occupation Classification (SOC) codes in the American Community Survey 2010 Public Use Microdata Sample Files were used to identify STEM occupations: 191010, 191020, 191030, 1910XX (Life Scientists); 192010, 192021, 192030, 192040, 192099 (Physical Scientists); 172011, 172041, 172051, 172061, 172070, 172081, 1720XX, 172110, 172121, 172131, 172141, 1721XX, 1721YY (Engineers); 151111, 151121, 151122, 151131, 151134, 15113X, 151141, 151142, 151143, 151150, 151199, 152011, 152031, 1520XX (Computer and Mathematical Occupations), 193011, 193030, 193051, 1930XX (Social Scientists); 113021, 119041, 119121 (STEM Managers); and 173020, 173031, 194011, 194021, 194031, 1940XX, 1940YY (STEM Technicians). Note that Computer and Mathematical Occupations include Mathematical Technicians and these technicians are excluded from the STEM Technician category. Also, the STEM Managers category includes architectural managers since they are included in SOC code 119041 (Architectural and Engineering Managers) and cannot be disentangled from engineering managers.

[14]The 4 percent figure equals the estimate of 2010 STEM employment in the defense industrial base divided by 2010 total U.S. STEM employment as detailed in Table 3-3.

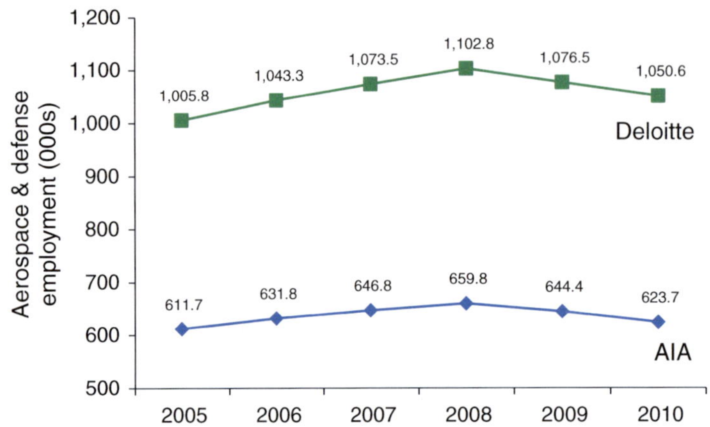

FIGURE 3-5 Aerospace and defense industry employment, 2005-2010.
SOURCE: Aerospace Industries Association (2012) and Deloitte (2012).

ment for the aerospace products and parts manufacturing industry, by occupational group. Table 3-5, in the annex, also contains employment information for specific occupations within an occupational group, when this information was available. Note that no employment information was available for two occupational groups—life sciences and social sciences. According to the Bureau of Labor Statistics' reporting standards for these estimates, these are instances in which the group has fewer than 50 jobs, the data are confidential, or the quality of the data is too poor to report. The "percent of industry" columns in Tables 3-4 and 3-5 show occupational employment as a percent of total industry employment and are useful for understanding which occupations are or are projected to be the most common in the aerospace products and parts manufacturing industry. The "percent of occupation" columns in the

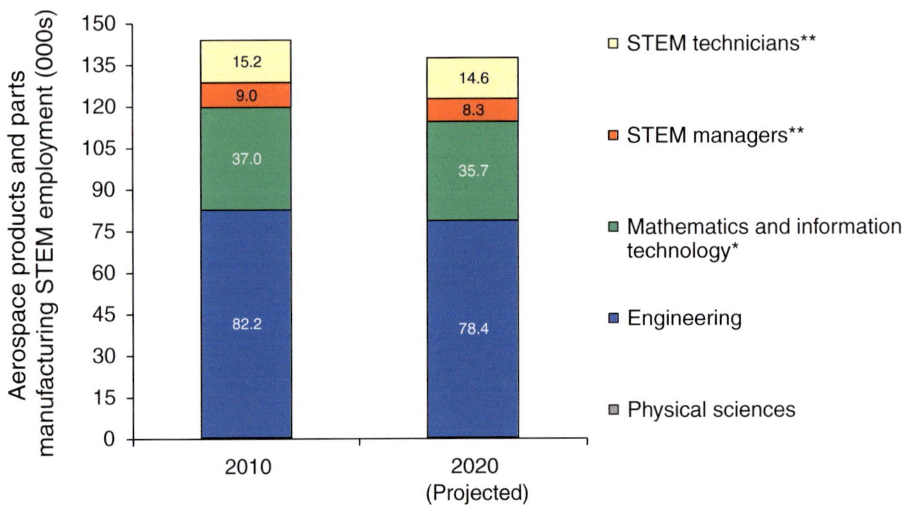

FIGURE 3-6 Aerospace products and parts manufacturing (NAICS 3364) STEM employment by occupational group, 2010 and 2020 (projected).
*A small number of Mathematical Technicians may be included in these figures.
**The information presented has been computed by the National Research Council.
SOURCE: Data are from the Bureau of Labor Statistics, U.S. Department of Labor, Employment Projections Program (www.bls.gov/emp/). No employment information is available for life sciences and social sciences because the groups had fewer than 50 jobs, or the data were confidential or of poor quality.

tables show occupational employment in the aerospace products and parts manufacturing industry as a percent of total occupational employment (across industries) and are useful for understanding how much of a specific occupation is concentrated in the aerospace products and parts manufacturing industry. As discussed earlier in the chapter, employment is very difficult to predict, so the projections presented here should be interpreted with caution.

Based on these Bureau of Labor Statistics estimates, 2010 employment in the aerospace products and parts manufacturing industry is approximately 477,000. Of this, STEM employment is 143,900, representing roughly 30 percent of total industry employment. Employment in the industry is projected to decline to approximately 462,000 by 2020, while STEM employment in the industry is projected to decline to 137,400. The projected decline in STEM employment is nearly proportional to the projected decline in industry employment such that STEM employment is expected to remain close to 30 percent of total industry employment. These results suggest that 3 out of every 10 jobs in the aerospace products and parts manufacturing industry is a STEM job, and this relationship is projected to remain until 2020. While STEM jobs are a major component of the aerospace products and parts manufacturing industry, these jobs represent 2 percent of total 2010 STEM employment (across industries). This figure is projected to decline to 1.6 percent by 2020. These figures suggest that only a small fraction of STEM workers are employed in the aerospace products and parts manufacturing industry.

Across the seven STEM occupational groups in the aerospace products and parts manufacturing industry, the largest is engineering, which accounts for 57 percent of 2010 industry STEM employment. This is followed by mathematics and information technology, which constitutes close to 26 percent of 2010 industry STEM employment. Employment in these two occupational groups is projected to decline by 2020, although these groups are expected to remain the largest components of STEM in the aerospace products and parts manufacturing industry. Looking again at Table 3-5, in the annex, the largest STEM occupation in the industry is aerospace engineers, with 2010 employment of 28,500, which represents 6 percent of total industry employment. By 2020, employment in this occupation is expected to fall to 25,500, representing 5.5 percent of projected industry employment. As of 2010, the aerospace engineering jobs in this industry equal slightly more than 35 percent of all U.S. aerospace engineering jobs. This figure is expected to decline to 30 percent by 2020. These results illustrate that a substantial proportion of aerospace engineers are employed in the aerospace products and parts manufacturing industry. Thus, events that affect this industry can be expected to have a disproportionate impact on aerospace engineers. Other relatively large STEM occupations in the industry include industrial engineers (16,000); software developers, systems software (15,000); other engineers (11,400), which could include salvage engineers, photonics engineers, ordnance engineers, and optical engineers; and mechanical engineers (11,300). However, these occupations are less concentrated in the aerospace products and parts manufacturing industry than are aerospace engineers.

Observation 3-4. While STEM jobs are a major component of the defense industrial base (approximately 3 in 10 jobs), these jobs represent a small fraction of total U.S. STEM employment (2-4 percent). A notable exception is aerospace engineers, a substantial proportion of whom are employed in the aerospace and defense industry.

The aforementioned American Community Survey data can also be used to examine the educational attainment and age of the STEM workforce in the defense industrial base. Based on this data, more than 70 percent of the 2010 defense industrial base STEM workforce is estimated to have a bachelor's degree (45 percent) or a master's degree (26 percent). Approximately 25 percent have less than a bachelor's degree, and relatively few have a doctoral degree (3.4 percent) or a professional degree (less 0.5 percent). The occupational group with the greatest proportion of the workforce with less than a bachelor's degree is STEM technicians (86 percent), distantly followed by mathematics and information technology (26 percent). For those in the defense industrial base STEM workforce who have a bachelor's degree or above, the most common fields of study are electrical engineering, mechanical engineering, general engineering, aerospace engineering, and computer science.[15]

Looking at the age distribution in the defense industrial base STEM workforce, Figure 3-7 shows the age distribution of this workforce in 2005 and 2011 (also see Table 3-6, in the annex). The defense industrial base STEM workforce has aged since 2005. The percentage of the workforce 55 and older increased from 18.6 percent

[15]Field of study reflects the major associated with an individual's bachelor's degree.

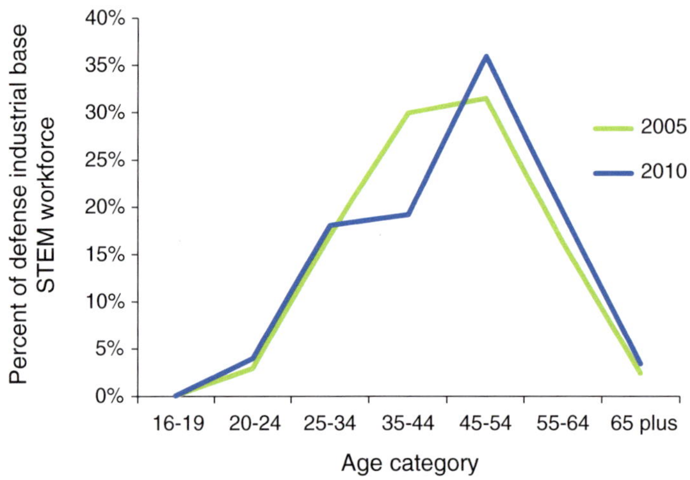

FIGURE 3-7 Age distribution of defense industrial base STEM workforce.
SOURCE: Data are from the American Community Survey Public Use Microdata Sample Files, 2005 and 2010. Tabulations by the National Research Council.

in 2005 to 22.8 percent in 2010. Moreover, the percentage of the workforce age 35-54 years old fell to 55.1 percent in 2010 from 61.4 percent in 2005. Despite an increase in the percentage of the workforce under the age of 35 from 20 percent in 2005 to 22.1 percent in 2010, this increase was not enough to prevent the median age from increasing from 45 in 2005 to 47 in 2010.

Observation 3-5. Approximately 75 percent of the defense industrial base STEM workforce has a bachelor's degree or above. Those with less than a bachelor's degree are concentrated in the STEM technician occupational group.

Observation 3-6. Despite an increase in the percentage of the defense industrial base STEM workforce that is under the age of 35, this increase has not been enough to prevent median age from increasing to 47 in 2010 (from 45 in 2005).

Additional information on the defense industrial base workforce is available in the 2011 edition of the annual Aviation Week Workforce Study (Hedden, 2011). The corporate study compiles demographic, hiring, and other information on the aerospace and defense industry workforce. According to the report, more than 30 aerospace and defense companies participated in the study, representing approximately 90 percent of the aerospace and defense workforce. Based on employment figures provided in the report, this 90 percent translates into employment of about 562,000 in 2010.

Figure 3-8 illustrates the age distribution of the aerospace and defense workforce in 2010. According to the chart, more than 50 percent of the workforce is between 36 and 55 years old, with roughly equal percentages younger than 36 and older than 55. The study also examined the age distribution by company size as measured by employee headcount. The age distributions by company size are similar to the age structure for the overall aerospace and defense workforce, with the exception of companies with fewer than 1,000 employees: these companies have more employees age 35 and under (33 percent) and fewer employees age 56 and older (17 percent). The study also notes that the "double-hump" in the industry's age distribution that was seen 5 years ago appears to have been corrected, resulting in a "smoother curve that allows for more active management of workforce structure in the future versus the previous trend of managing to retirement alone" (Hedden, 2011, p. 14).

The Aviation Week study also captures information on retirements and retirement eligibility in the aerospace and defense industry.[16] Figure 3-9 shows 2010 retirement rates and retirement eligibility for the aerospace and

[16]According to the Aviation Week study, all companies included in the study except two used an age threshold of 62 to identify those eligible to retire.

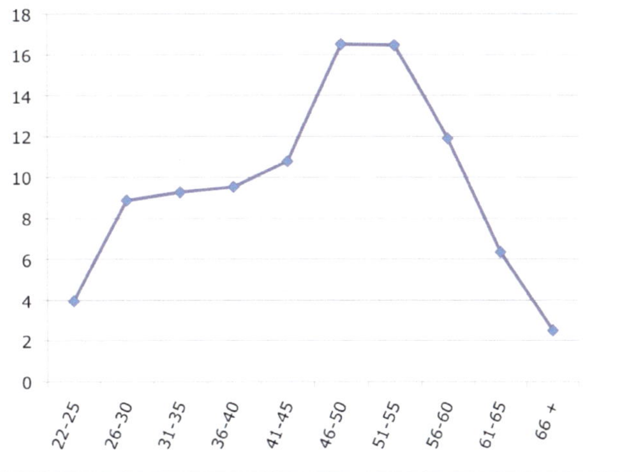

FIGURE 3-8 Age distribution in the aerospace and defense industry workforce.
SOURCE: Hedden (2011).

defense industry workforce by job category as well as projected retirement eligibility in 2014 (also see Table 3-7, in the annex). In 2010, 13.8 percent of employees were eligible to retire, but only a relatively small percentage of employees actually retired (1.2 percent). The results are similar across job categories such that only a small percentage of employees in a given job category retired relative to the percent in the job category who were eligible. The Aviation Week study notes that although "the economy had an impact on ability to retire, Human Resources leaders also believe employees are choosing to remain on the job longer" (Hedden, 2011, p. 12), suggesting that the low observed retirement rates may be due to a combination of the recent recession and other factors. The study further notes that the "real proof will come in the level of retirements as the economy rebounds" (Hedden, 2011, p. 12). The lowest retirement eligibility in 2010 was in the software development job category (5.9 percent), while the highest was in the non-exempt job category (15.4 percent). By 2014, overall retirement eligibility is expected to rise to 22.8 percent, with the test and evaluation and supply chain job categories expected to both reach 29.4 percent.

Looking at retirement eligibility by company size (see Table 3-8, in the annex) shows that current and projected eligibility rates are similar to the overall figures for companies of differing sizes, except for those with fewer than 1,000 employees, in which eligibility rates are considerably lower. For example, in 2010 only 2.8 percent of employees in these small companies were eligible to retire; this is expected to increase only to 3.2 percent by 2016.

Observation 3-7. As of 2010, about 15 percent of the aerospace and defense workforce was estimated to be eligible to retire, but less than 2 percent of employees actually retired. This low observed retirement rate, however, may be due (at least in part) to the recent recession, and retirements may increase as the economy improves.

In terms of hiring in the aerospace and defense industry, the companies included in the Aviation Week study reported that they planned to hire close to 32,000 people in 2011, which represents about 5.7 percent of the 2010 aerospace and defense workforce in these companies. The hiring estimates for 2012 and beyond are lower: approximately 22,000 in 2012 and in 2013 and roughly 14,000 in 2016. Companies in the study reported that the three most difficult to fill positions are in systems engineering, aerospace engineering, and mechanical engineering, although the most difficult to fill position varies by company size. Larger companies with 10,000 or more employees reported systems engineering positions as the most difficult to fill; medium-sized companies with fewer than 10,000 employees but more than 1,000 employees reported mechanical engineering positions; and small companies with fewer than 1,000 employees reported aerospace engineering positions as the most difficult to fill.

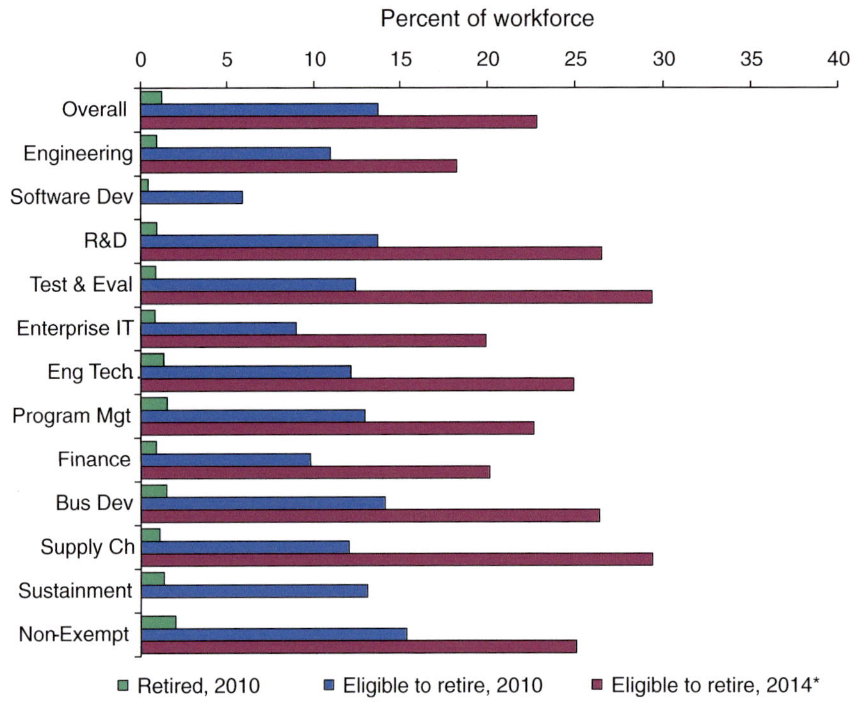

FIGURE 3-9 Retirements and retirement eligibility for the aerospace and defense industry workforce by job category, 2010 and 2014.
*Projected.
SOURCE: Data are from Hedden (2011). The projected percent eligible for retirement in 2014 is not available for software development and for sustainment.

Observation 3-8. Aerospace and defense companies report that the three most difficult to fill positions are in systems engineering, aerospace engineering, and mechanical engineering.

STEM WORKFORCE IN DOD

The length of time required for the committee to gain access to data on the DOD STEM workforce limited the scope of the analysis that was possible within the timeframe of the study. To examine the STEM workforce within DOD, the committee relied on information from two primary sources. The first source was individual-level personnel records provided by the Defense Manpower Data Center (DMDC). The information provided included records for both civilian and military DOD personnel. The second source was FedScope, an online tool that enables users to generate information on the federal civilian workforce. This latter database was used to fill in gaps in the data received from DMDC.

In requesting DOD personnel records, the committee provided DMDC with a list of STEM occupations to include in the data files. This list was based on occupations used in a 2008 National Science Foundation study of federal civilian scientists and engineers (National Science Foundation and Division of Science Resources Statistics, 2008).[17] The committee realizes this may not be an appropriate definition of STEM for DOD. However, creating a definition of STEM for DOD was outside the scope of the committee's task and resources. Furthermore, the

[17]Note that the National Science Foundation uses the term "scientists and engineers" (S&E) for its reporting of information related to science, technology, engineering, and mathematics.

committee is aware that the Department of the Air Force has an official definition of STEM that can be found in *Bright Horizons*, the Air Force STEM Strategic Roadmap.

DOD Civilian STEM Workforce

For its study of the federal scientist and engineering workforce, NSF identified scientist and engineering occupations using 80+ Office of Personnel Management (OPM) occupational series codes. Based on this list, DMDC provided the committee with the following information:

- Records for DOD personnel employed in these OPM occupational series codes as of the end of each of the 2001-2011 fiscal years[18] (i.e., annual snapshots);
- Records for DOD personnel entering these OPM occupational series codes during the 2001-2011 fiscal years from events such as appointments, transfers, conversions, realignments, and reassignments (i.e., annual entries);
- Records for DOD personnel exiting these OPM occupational series codes during the 2001-2011 fiscal years from events such as separations from the federal civil service, transfers, conversions, realignments, and reassignments (i.e., annual exits).

The NSF study groups these 80+ OPM occupational series codes into five major occupational groups—computer and mathematical scientist, engineer, life scientist, physical scientist, and social scientist—and these major occupational groups are broken down into 19 minor occupational groups.[19] Table 3-9, in the annex, shows the crosswalk between major and minor STEM occupational groups and OPM occupational series codes, as it has been applied to the study at hand.[20]

Using information provided by DMDC and supplemented by FedScope, the committee examined the size and composition of the DOD civilian STEM workforce.[21] Over the past 5 years there has been a slight increase in the percent of total DOD civilian employment that is STEM, increasing from 18.4 percent in 2007 to 19.4 percent in 2011 (see Figure 3-10). These results suggest that roughly 1 in 5 DOD civilian employees is in a STEM occupation. Moreover, more than 46 percent of 2011 federal civilian STEM employees are in DOD, and this figure has increased modestly since 2007 (see Figure 3-11). Looking at STEM occupational groups, more than two-thirds of federal civilian engineers are in DOD and close to 50 percent of federal civilian computer and mathematical scientists are in DOD. Slightly more than a quarter of federal civilian physical scientists and social scientists are in DOD. On the other end of the spectrum, less than 15 percent of federal civilian life scientists are in DOD. These results suggest that a large fraction of STEM talent in the federal civilian workforce, most notably engineers and computer and mathematical scientists, is employed by DOD.

Observation 3-9. Roughly 1 in 5 DOD civilian employees is in a STEM occupation, and close to half of the federal civilian STEM workforce and more than two-thirds of federal civilian engineers are in the DOD.

Total DOD civilian STEM employment is roughly 151,000 (see Figure 3-12). This represents about 2.3 percent of total U.S. STEM employment (excluding technicians), suggesting that only a small fraction of U.S. STEM workers are employed by DOD.[22] The largest occupational groups are engineers (71,123) and computer

---

[18]The federal government's fiscal year begins on October 1 and runs until September 30. For example, the 2011 fiscal year ran from October 1, 2010, to September 30, 2011.

[19]NSF's study of federal scientists and engineers excludes technicians in the federal civil service. In keeping with this definition, the current study also excludes DOD civilian STEM technicians. According to data available in FedScope, there are approximately 23,000 civilian STEM technicians in the DOD.

[20]For a more detailed description of the OPM occupational series codes included in the analysis, see the Office of Personnel Management's *Handbook of Occupational Groups and Families*, May 2009 (available at http://www.opm.gov/fedclass/gshbkocc.pdf).

[21]Note that in conducting this examination, the committee did not impose any minimum education requirement (e.g., bachelor's degree and above) in identifying STEM employees. Rather, STEM employees were identified solely based on their occupation.

[22]The 2.3 percent figure equals 2011 DOD civilian STEM employment divided by 2010 total U.S. STEM employment (excluding technicians) as detailed in Table 3-3.

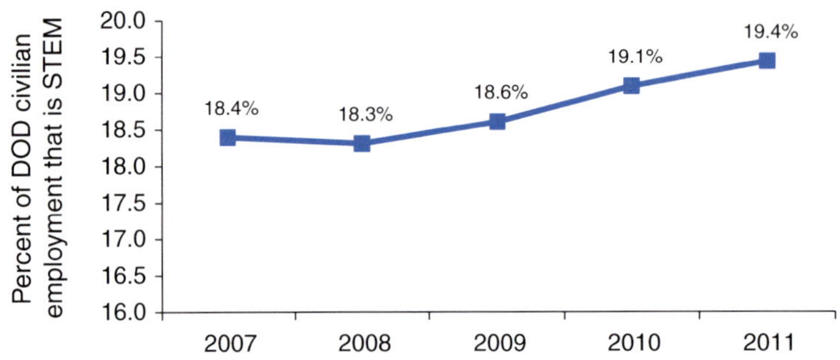

FIGURE 3-10 Department of Defense civilian STEM employment as a percent of total DOD civilian employment, 2007-2011.
NOTE: Figures are as of the fiscal year-end (e.g., fiscal year 2011 is as of September 30, 2011).
SOURCE: Data are from FedScope. Tabulations by the National Research Council.

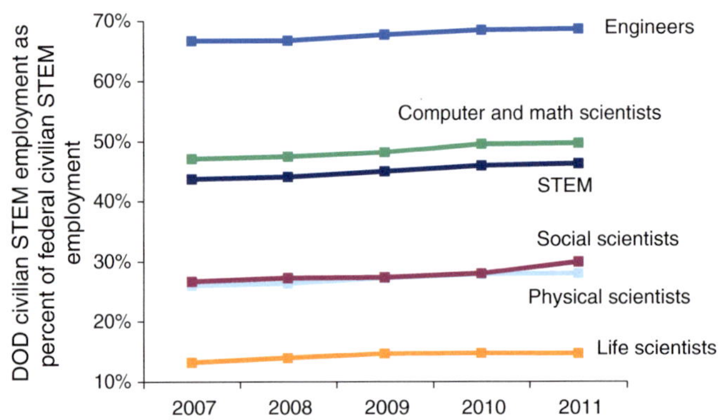

FIGURE 3-11 Department of Defense civilian STEM employment as a percent of federal civilian STEM employment by major occupational group, 2007-2011.
NOTE: Figures are as of the fiscal year-end (e.g., fiscal year 2011 is as of September 30, 2011). Federal civilian STEM employment is based on the agencies included in FedScope.
SOURCE: Data are from FedScope. Tabulations by the National Research Council.

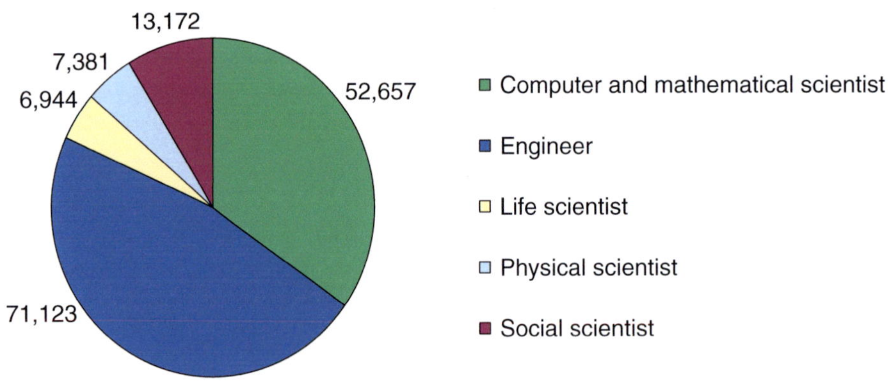

FIGURE 3-12 Department of Defense civilian STEM employment by major occupational group, 2011.
NOTE: Figures are as of the fiscal year-end (September 30, 2011).
SOURCE: Data provided by the Defense Manpower Data Center. Tabulations by the National Research Council.

and mathematical scientists (52,657). There are relatively few life and physical scientists (less than 5 percent each) and about 9 percent are social scientists. These results are similar to those for the STEM workforce in the defense industrial base discussed earlier. In fact, for both the civilian STEM workforce in the DOD and the STEM workforce in the aerospace products and parts manufacturing industry, more than 80 percent of the workforce is engineers and computer and mathematical scientists, illustrating the importance of these fields both inside DOD and in the organizations that support DOD. For 2011 DOD civilian STEM employment by OPM occupational series code, see Table 3-9, in the annex.

Across all occupational groups, DOD civilian STEM employment has increased since 2001 (Figure 3-13; see also Table 3-10, in the annex). The highest growth rate has been in the social scientist occupational group, which experienced an annual growth rate of 6.5 percent over the period 2001-2011. Physical scientists and engineers experienced the lowest annual growth rates over this period (0.9 percent and 2.2 percent, respectively). Due to these lower growth rates, the share of the DOD civilian STEM workforce that is engineers and physical scientists has declined slightly over the past 10 years. DOD representatives almost uniformly state that they foresee no shortage of STEM workers in the years ahead except in a few specialty fields (National Research Council, 2012).

Observation 3-10. Total DOD civilian STEM employment is approximately 150,000, where 47 percent are in engineering and 35 percent are in computer and mathematical scientist occupations; however, the DOD civilian STEM workforce represents only a small fraction of the total U.S. STEM workforce (approximately 2 percent).

For the analysis included in this chapter, there are more than 80 OPM occupational series included as STEM occupations. Of these, the 20 largest account for more than 90 percent of 2011 DOD civilian STEM employment, with the top 5 occupations accounting for more than 50 percent. Table 3-11, in the annex, lists these top 20 occupations. The largest occupation is 2210-Information Technology Management, which accounts for close to 24 percent of 2011 DOD civilian STEM employment. According to the OPM's *Handbook of Occupational Groups and Families*, these positions "manage, supervise, lead, administer, develop, deliver, and support information technology (IT) systems and services . . . for which the paramount requirement is knowledge of IT principles, concepts, and methods; e.g., data storage, software applications, networking" (Office of Personnel Management, 2009, p. 120). The four next-largest occupations are all in engineering: 0855-Electronics Engineering, 0801-General Engineering, 0830-Mechanical Engineering, and 0810-Civil Engineering. The fifth largest is 0132-Intelligence, which is part of the social scientist occupational group. These positions "require a basic knowledge and understanding of one or

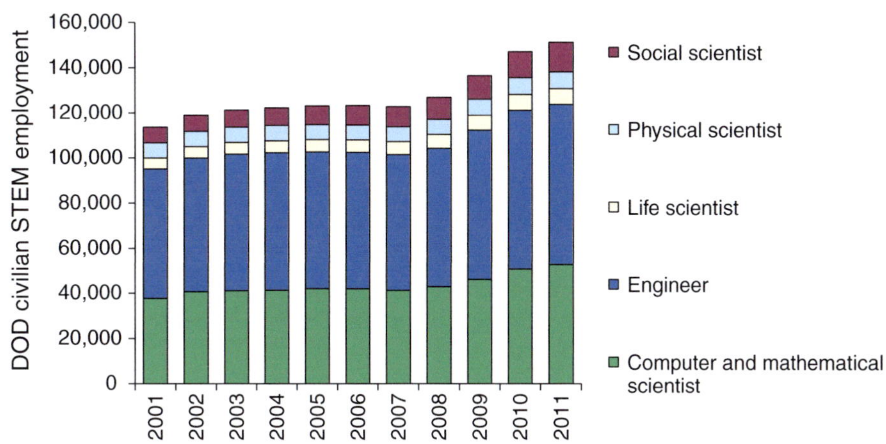

FIGURE 3-13 Department of Defense civilian STEM employment by major occupational group, 2001-2011.
NOTE: Figures are as of the fiscal year-end (e.g., fiscal year 2011 is as of September 30, 2011).
SOURCE: Data provided by the Defense Manpower Data Center. Tabulations by the National Research Council.

more of the natural or social sciences, engineering, or military science, but do not demand, as a primary qualification requirement, full knowledge of the current state of the art" (Office of Personnel Management, 2009, p. 26).

The OPM STEM occupational series with the highest growth rates over the period 2001-2011 are listed in Table 3-12, in the annex. The fastest growing occupation is 0130-Foreign Affairs with an annual growth rate of 11.3 percent, followed closely by 0150-Geography with an annual growth rate of 10.6 percent. Both of these occupations are in the social scientist occupational group. While social scientists make up less than 9 percent of the DOD civilian STEM workforce, 7 of the 20 fastest growing occupations are in the social scientist occupational group. Looking at the other occupational groups, the fastest growing occupation in the computer and mathematical scientist occupational group is 1550-Computer Science (6 percent annual growth); the fastest growing in the engineer occupational group is 0858-Bioengineering & Biomedical Engineering (6.6 percent annual growth); the fastest growing in the physical scientist occupational group is 1306-Health Physics (4.1 percent annual growth); and the fastest growing in the life scientist occupational group is 0401-General Natural Resources Management and Biological Sciences (7.9 percent annual growth). A complete list of employment over the period 2001-2011 and growth rates by OPM occupational series are provided in Table 3-13, in the annex.

Observation 3-11. The largest DOD civilian STEM occupations are information technology management and electronics engineering, which together number more than 50,000 employees.

The size and composition of the DOD STEM civilian workforce vary across the three services—Department of the Air Force, Department of the Army, Department of the Navy—and the Department of Defense Agencies[23] (Figure 3-14; see also Table 3-14, in the annex). Looking at 2011 employment, the Department of the Army has the largest number of civilian STEM employees (55,760), which represents about 37 percent of the total DOD STEM civilian workforce. The Department of the Navy is close behind with almost 53,000 employees, while the Department of the Air Force has slightly less than 30,000 STEM employees. The smallest STEM employment is in the Department of Defense Agencies (12,855). In the Departments of the Air Force, Army, and Navy, engineers constitute the largest STEM occupational group, while computer and mathematical scientists constitute the largest STEM occupational group in the Department of Defense Agencies.

Observation 3-12. The largest STEM occupational group in the Departments of the Air Force, Army, and Navy is engineers; however, in the Department of Defense Agencies, computer and mathematical scientists represent the largest STEM occupational group.

Since this committee's study did not impose any minimum education requirement in identifying DOD STEM employees, the committee felt it was important to examine the educational attainment of the DOD civilian STEM workforce.[24] One thing to keep in mind, however, is that educational attainment information is typically captured at the time of appointment and may not be updated in the human resource information system if an employee subsequently earns a higher-level degree.[25] Thus, the educational attainment information presented here should be viewed with this caveat.

Figure 3-15 shows highest degree attained for the DOD civilian STEM workforce over the period 2001-2011. The results suggest that the distribution of educational attainment has been stable over the past 10 years. As of 2011, approximately 48 percent have a bachelor's degree, close to a quarter have a master's degree, and roughly 5 percent have a doctoral degree. Somewhat surprisingly, about 22 percent have less than a bachelor's degree. Looking at Figure 3-16, however, shows that educational attainment varies by STEM occupational group. The relatively large percentage of people with less than a bachelor's degree primarily reflects the large percentage of computer

[23]Department of Defense Agencies includes agencies such as the Defense Logistics Agency, the Defense Information Systems Agency, the Missile Defense Agency, and the Defense Contract Management Agency. Employees in the Marine Corps have been included with the Department of the Navy.

[24]See Table 3-15, in the annex, for a crosswalk between highest degree attained and OPM's classification of educational attainment.

[25]FedScope, About Our Data (EHRI-SDM), Data Element Information (available at http://www.fedscope.opm.gov/datadefn/aehri_sdm.asp#cpdf6).

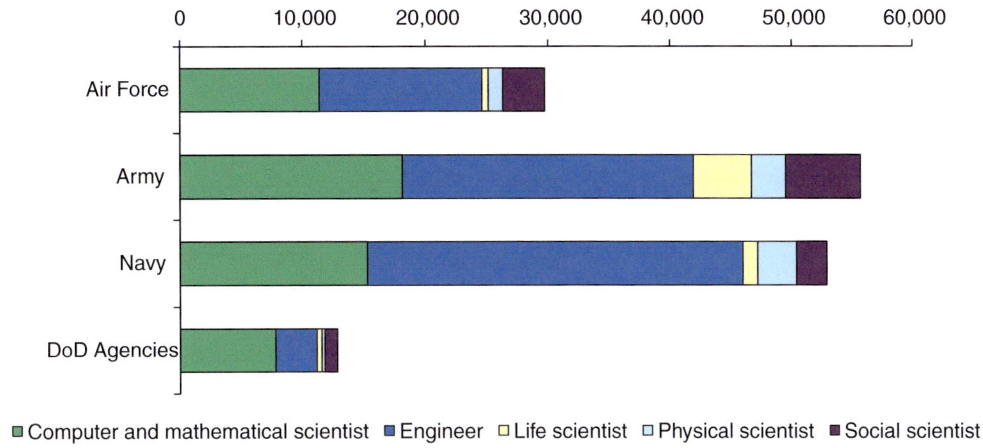

FIGURE 3-14 Department of Defense civilian STEM employment by department and major occupational group, 2011.
NOTE: Figures are as of the fiscal year-end (September 30, 2011).
SOURCE: Data provided by the Defense Manpower Data Center. Tabulations by the National Research Council.

and mathematical scientists with less than a bachelor's degree (49.4 percent). The life scientist and social scientist occupational groups also have a sizable percentage of employees with less than a bachelor's degree (21.5 percent and 31.7 percent, respectively), although these groups also have a substantial proportion with a graduate degree (32.9 percent and 41.3 percent, respectively). Relatively few in the engineering and physical scientist occupational groups have less than a bachelor's degree (2.2 percent and 5.5 percent, respectively), with a majority of those in engineering having a bachelor's degree and a majority of those in the physical sciences having a graduate degree.

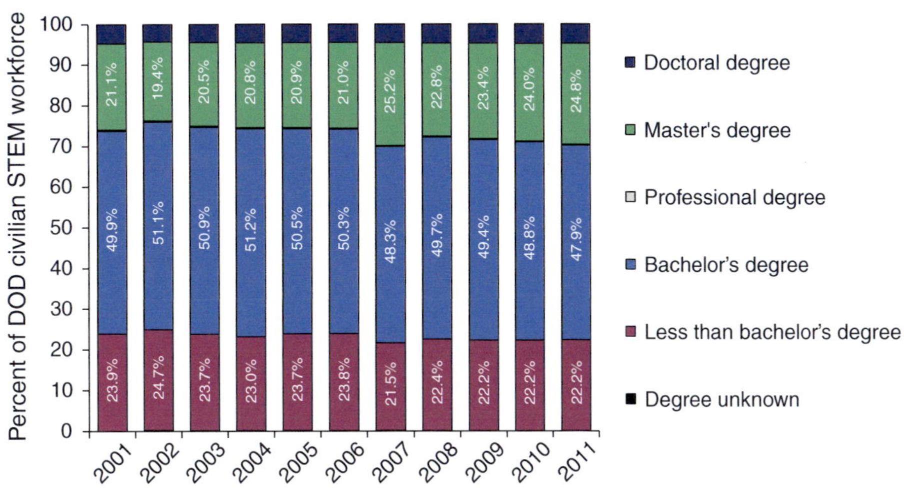

FIGURE 3-15 Highest degree attained for Department of Defense civilian STEM workforce, 2001-2011.
NOTE: Figures are as of the fiscal year-end (September 30, 2011).
SOURCE: Data provided by the Defense Manpower Data Center. Tabulations by the National Research Council.

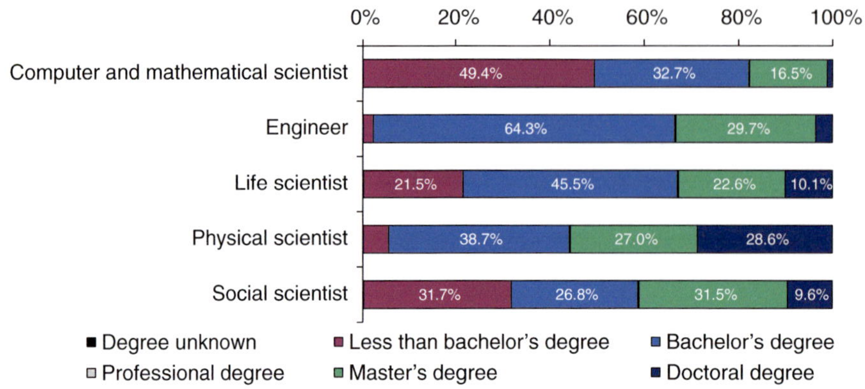

FIGURE 3-16 Highest degree attained for Department of Defense civilian STEM workforce by major occupational group, 2011.
NOTE: Figures are as of the fiscal year-end (September 30, 2011).
SOURCE: Data provided by the Defense Manpower Data Center. Tabulations by the National Research Council.

Observation 3-13. Roughly half of the DOD civilian STEM workforce have at most a bachelor's degree, and close to 30 percent have a graduate degree; the remaining one-fifth, with less than a bachelor's degree, are concentrated in computer and mathematical scientist occupations.

The committee also examined fields of study for those with a postsecondary degree (i.e., those with an associate's degree or higher). In the data provided to the committee by DMDC, degree field is captured using the Classification of Instructional Programs (CIP), a taxonomy of instructional programs developed and maintained by the U.S. Department of Education's National Center for Education Statistics (NCES). The taxonomy has at its foundation a series of six-digit codes that represent specific programs of instruction. For example, the six-digit code for Computer Hardware Engineering is 14.0902; the six-digit code for Computer Software Engineering is 14.0903. Six-digit codes can be aggregated into four-digit codes representing groupings of six-digit programs that share similar content. Continuing on with the earlier example, programs 14.0902 and 14.0903 are in the four-digit CIP code 14.09, labeled Computer Engineering. In turn, four-digit codes can be rolled up into two-digits CIP codes, representing the broadest grouping of related programs. The 14.09 CIP code is part of the two-digit CIP code 14, labeled Engineering.

Table 3-16, in the annex, lists the most common fields of study (using four-digit CIP codes) as of 2011 for those in the DOD civilian STEM workforce with a postsecondary degree. The three most common fields of study are in engineering—electrical, electronics and communications engineering (CIP 14.10); mechanical engineering (CIP 14.19); and civil engineering (CIP 14.08)—and more than 30 percent of DOD civilian STEM employees with a postsecondary degree have their highest degree in one of these three fields. Moreover, more than 50 percent of DOD civilian STEM employees who have a postsecondary degree have a degree in an engineering (CIP 14) or engineering technologies/technician field (CIP 15) (see Table 3-17, in the annex). After engineering, the most common degree field is in the area of computer and information sciences and support services (CIP 11), most notably computer and information sciences, general (CIP 11.01); computer science (CIP 11.07); and information science/studies (CIP 11.04). Interestingly, almost 10 percent of those with a postsecondary degree have their highest degree in the area of business, management, marketing, and related support services (CIP 52). Looking at field of study by major occupational group (see Table 3-18, in the annex) shows that close to 22 percent of employees in the computer and mathematical scientist occupational group with a postsecondary degree, and more than 13 percent of employees in the social scientist occupational group with a postsecondary degree, have a degree in this field.

Observation 3-14. More than 50 percent of the DOD civilian STEM workforce with a postsecondary degree has a degree in an engineering or engineering technologies/technician field, the most common being electrical, electronics, and communications engineering; mechanical engineering; and civil engineering.

To examine concerns about the prospect of looming retirements among the DOD civilian STEM workforce, the committee examined the age distribution of the workforce. Over the past 10 years median age has increased from 45 in 2001 to 47 in 2011 (Figure 3-17; see also Table 3-19, in the annex). However, average age has remained stable over this period and stands at 45.3 as of 2011, compared to 45.6 in 2001. The stability in average age is due to the relatively large number of people in the workforce under age 36 in 2011 (24.4 percent) as compared to 2001 (14.0 percent). Over time, as older employees retire, this younger cohort will become more dominant and will likely result in a more desirable age profile for the DOD civilian STEM workforce. Comparing Figure 3-17 to Figure 3-8, the most recent age distribution in the STEM civilian workforce is strikingly similar to the age distribution in the aerospace and defense industry.

Examining the most recent age distribution of the DOD civilian STEM workforce by occupational group shows that the oldest group is physical scientists, while the youngest group is engineers (Figure 3-18; see also Table 3-20, in the annex). Less than 20 percent of the physical scientist workforce is 35 and under, compared to 30 percent of engineers. Moreover, more than 27 percent of the physical scientist workforce is older than 55, compared to 16.5 percent of engineers.

Observation 3-15. Although the DOD civilian STEM workforce has aged over the past 10 years (in terms of median age), there is a relatively large cohort of people under age 36, which may result in an improvement in the workforce's age profile as older employees retire. Moreover, those in physical scientist occupations are generally the oldest and those in engineering occupations are generally the youngest.

A more direct indicator of the possibility of retirement among the DOD civilian STEM workforce is the percentage of the workforce that is eligible to retire. The retirement eligibility of a civilian government employee

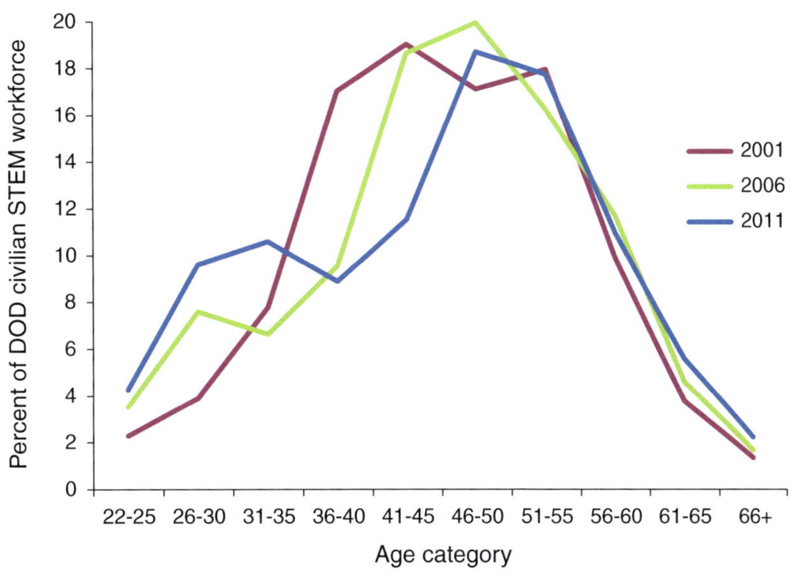

FIGURE 3-17 Age distribution of Department of Defense civilian STEM workforce, selected years: 2001, 2006, and 2011.
NOTE: Figures are as of the fiscal year-end (e.g., September 30, 2011).
SOURCE: Data provided by the Defense Manpower Data Center. Tabulations by the National Research Council.

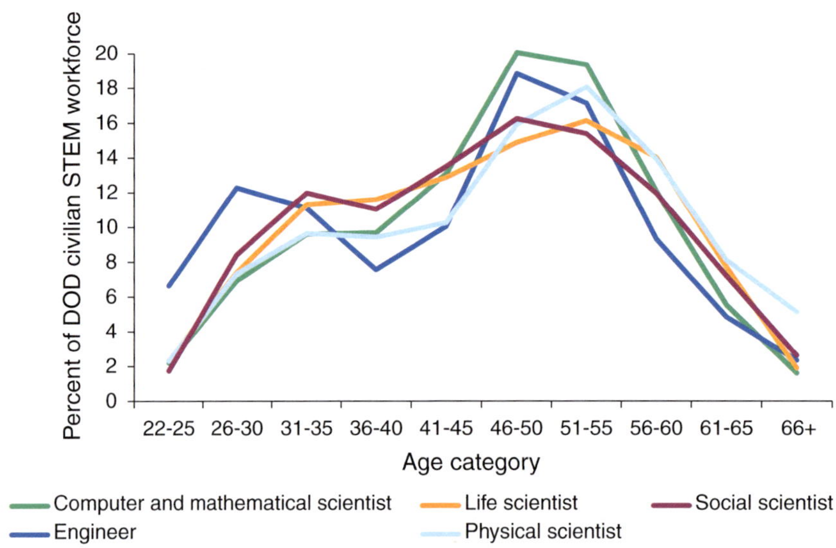

FIGURE 3-18 Age distribution of Department of Defense civilian STEM workforce by major occupational group, 2011.
NOTE: Figures are as of the fiscal year-end (e.g., September 30, 2011).
SOURCE: Data provided by the Defense Manpower Data Center. Tabulations by the National Research Council.

is determined by the retirement system the employee is subject to—the Civil Service Retirement System (CSRS) or the Federal Employees Retirement System (FERS). Under both systems, an employee can qualify for regular retirement as young as age 55. For example, an employee subject to CSRS can qualify for regular retirement at age 55 if the employee has 30 or more years of creditable service. Under FERS, an employee can qualify for retirement at age 55 if the employee has 30 years or more of creditable service and was born prior to 1948 (Office of Personnel Management, 2012). The percent eligible for retirement has remained relatively stable over the past 10 years and stands at 32.3 percent as of 2011 (see Figure 3-19). This is down slightly from a high of 35.5 percent in 2007.

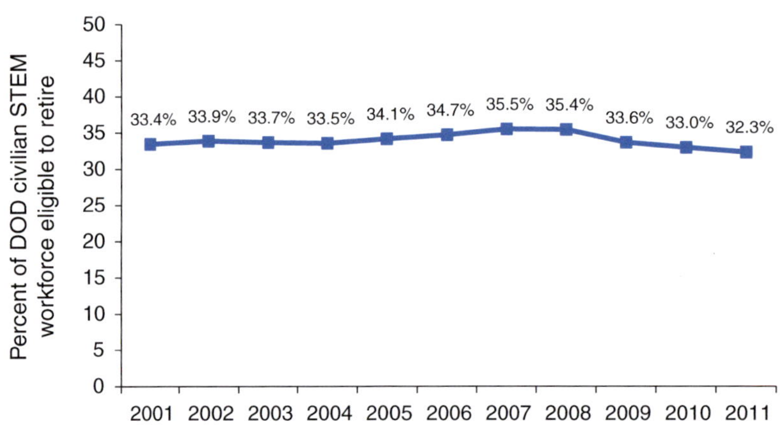

FIGURE 3-19 Retirement eligibility of Department of Defense civilian STEM workforce, 2001-2011.
NOTE: Figures are as of the fiscal year-end (September 30, 2011).
SOURCE: Data provided by the Defense Manpower Data Center. Tabulations by the National Research Council.

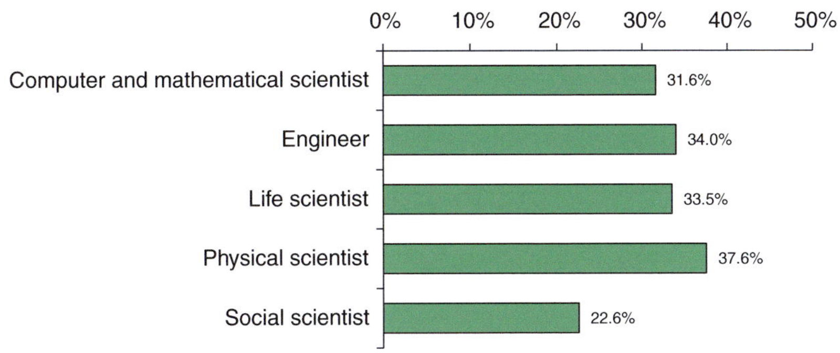

FIGURE 3-20 Retirement eligibility of Department of Defense civilian STEM workforce by major occupational group, 2011.
NOTE: Percentages are as of the fiscal year-end (September 30, 2011).
SOURCE: Data provided by the Defense Manpower Data Center. Tabulations by the National Research Council.

Looking at 2011 retirement eligibility by major occupational group (Figure 3-20) shows that the highest eligibility is in the physical scientist group (37.6 percent) and the lowest is in the social scientist group (22.6 percent).

Although close to a third of the DOD civilian STEM workforce is eligible to retire, retirement rates are relatively low (see Figure 3-21). The most recent retirement rate for the workforce is 2.8 percent, suggesting that relatively few of those eligible to retire in 2011 actually retired. Even in the physical scientist occupational group, which has the highest retirement rate, the rate is less than 4 percent. Moreover, separation rates, reflecting various personnel actions that result in the exit of an employee from DOD, are relatively low. The 2011 separation rate for the DOD civilian STEM workforce is 7.4 percent. This figure includes quits, retirements, reductions in force, termination/removals, transfers, death, and other separation reasons. The lowest separation rate is in the engineer occupational group (5.5 percent); the highest is in the social scientist group (12.5 percent). These figures are low compared to the 2011 separation rate for the federal government as a whole (13.1 percent) and for the private sector

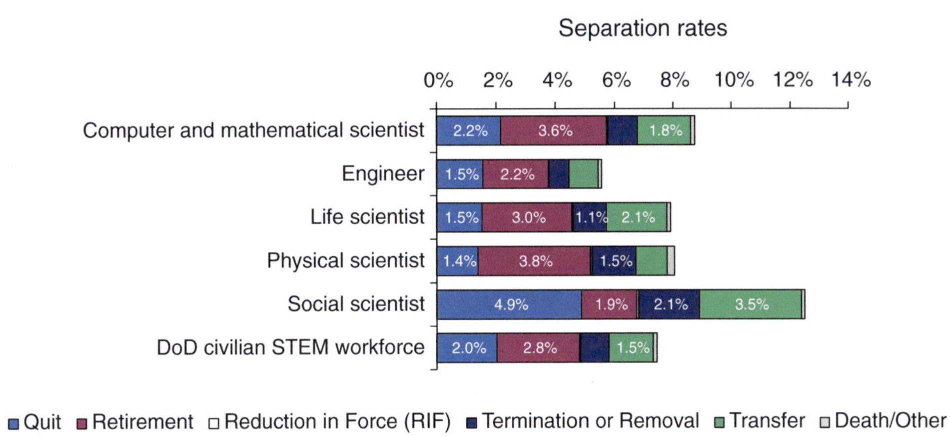

FIGURE 3-21 Department of Defense civilian STEM separation rates by type and major occupational group, 2011.
NOTE: Figures are for the 2011 fiscal year (October 1, 2010 through September 30, 2011). A separation is a personnel action resulting in the loss of an employee from an agency's staff. The separation rate equals the total number of separations of a given type during the 2011 fiscal year divided by average employment during the year. Average employment is defined as the average of employment at the end of the 2011 fiscal year and employment at the end of the 2010 fiscal year.
SOURCE: Data are from FedScope. Tabulations by the National Research Council.

(41 percent), both calculated from the Bureau of Labor Statistics' Job Openings and Labor Turnover Survey.[26] This latter finding that the separation rate for the federal workforce is substantially lower than the separation rate for the private sector does not appear to be an artifact of the recent recession. A similar pattern emerges from an examination of historical separation rates. For example, over the period 2002-2011, the average annual separation rate in the private sector was 2.4 times larger than the average annual separation rate in the federal government. Looking by type of separation shows that the largest difference is for voluntary turnover whereby the average annual voluntary turnover rate in the private sector over the period 2002-2011 was 3.8 times larger than the average annual voluntary turnover rate in the federal government over this same period.

Observation 3-16. Roughly one-third of the DOD civilian STEM workforce is eligible to retire; however, the actual rate of retirement is less than 4 percent.

DOD Military STEM Workforce

Compared to the DOD civilian data, the data the committee received on the DOD military workforce was considerably more problematic. The primary issue was the identification of military STEM occupations. In compiling the data, DMDC identified for the committee military STEM occupations using the DOD classification of occupations. The DOD classification is designed to group similar occupations across the military services into a single, consistent taxonomy for analytical purpose (Department of Defense Office of the Under Secretary for Personnel and Readiness, 2001). Some examples include Analysis; Chemical; Construction and Utilities; Data Processing; Intelligence, General; Physical Scientists; and Social Scientists (see Table 3-21, in the annex for a complete list of the DOD occupation codes identified as STEM by DMDC).

However, examining the service-specific occupations that fall into the DOD occupations identified as STEM, suggests that (1) some of the DOD occupations identified as STEM may not be STEM and (2) some of the DOD occupations identified as STEM contain a combination of STEM and non-STEM jobs. As an example of the former, it was not clear to the committee why those in the Manpower and Personnel DOD occupation code were considered to be STEM. A review of the service-specific occupations within this DOD occupation shows that the largest occupations are human resource officers. Another example is the Administrators, General DOD occupation, which contains primarily adjutants, executive officers, and administrative officers. Other DOD occupations appear to contain a combination of STEM and non-STEM jobs. For example, the Analysis DOD occupation contains a large number of signals intelligence analysts, which appears to be a STEM job, but also contains a large number of translator jobs, which is not a STEM job group. Moreover, the Educators and Instructors DOD occupation contains professors in STEM fields (e.g., engineering, physical science) and non-STEM fields (e.g., history, English).

After examining the data, the committee decided it did not have confidence in the usefulness of the DOD classification system to identify military STEM occupations. Thus, the committee did not feel justified in drawing any conclusions about the military STEM workforce from the data provided by DMDC.

Observation 3-17. Due to difficulties associated with defining military STEM occupations, the committee was not able to assess the military STEM workforce using the data provided by DMDC.

FINDINGS AND RECOMMENDATIONS

Finding 3-1. The defense industrial base has made good strides toward creating a process for generating and sharing information on the defense industrial base workforce, most notably through the annual Aviation Week Workforce Study, which is in association with the Aerospace Industries Association, the American Institute of Aeronautics and

[26]Bureau of Labor Statistics, Job Openings and Labor Turnover Survey (www.bls.gov/jlt/). For the survey, a separation includes quits, layoffs and discharges, retirements, transfers to other locations, deaths, or separations due to employee disability. It does not include transfers within a given establishment, employees on strike, employees of temporary help agencies, employee leasing companies, outside contractors, and consultants working at the sampled establishment (see www.bls.gov/jlt/jltdef.htm#4).

Astronautics, and the National Defense Industries Association. However, a comprehensive source of information on the defense industrial base workforce is not available. For example, the Aviation Week study excludes some key areas of the industry, such as ship building and repairing. While the Deloitte (2012) study (commissioned by the Aerospace Industries Association) uses a broader measure of the industry, the focus of the study is not on the workforce per se, but rather, the economic impact of the defense industrial base on the U.S. economy. Moreover, these studies do not specifically address the STEM workforce in the defense industrial base.

Recommendation 3-1. The DOD should form a working group with the defense industrial base, perhaps as an activity of the DOD's STEM Board of Directors, to develop a definitive, comprehensive survey of the defense industrial base workforce to facilitate the management of this workforce, forecast critical needs, and respond to workforce challenges as they relate to matters of national defense. Special consideration should be given to identifying and tracking the STEM workforce in this industry.

Finding 3-2. The Department of Defense has not defined a STEM taxonomy for both its DOD military members and civil service STEM personnel. Having an approved STEM taxonomy is an important first step in addressing the challenges facing the DOD in the management of its STEM workforce. Moreover, a single approved STEM definition would provide the basis for identifying, comparing, and tracking STEM expertise.

Recommendation 3-2a. The Department of Defense needs officially to define a STEM taxonomy that spans the military and civilian workforce in a manner that meets its requirements and accommodates the mission-driven needs of the services within the department. When determining whether to define STEM narrowly or more broadly, DOD needs to take into consideration the purposes for which this definition will be used and the funding issues addressed in Finding 4-1, giving due consideration to non-traditional STEM fields such as social sciences. Within the current budget environment the committee advises using a more narrowly defined STEM taxonomy for making training and education investment decisions for critical STEM skills.

Recommendation 3-2b. Premised on an officially defined DOD STEM taxonomy, the department needs to develop an analytical capability to manage this workforce, manage STEM workforce historical data, forecast critical needs, and respond to workforce challenges.

REFERENCES

Bureau of Labor Statistics. 2012. Employment Projections, Frequently Asked Questions. Available at http://www.bls.gov/emp/ep_faq_001.htm#account (accessed June 6, 2012).

Bureau of Labor Statistics. Undated. Standard Occupational Classification System.

Carnevale, A.P., N. Smith, and M. Melton. 2011. STEM: Science, Technology, Engineering and Mathematics. Washington, D.C.: Georgetown University.

Cover, B., J.I. Jones, and A. Watson. 2011. Science, technology, engineering, and mathematics (STEM) occupations: A visual essay. Monthly Labor Review 134(May):3-15.

CPST. 2006a. A Half-Century Snapshot of the STEM Workforce, 1950 to 2000. New York: Commission on Professionals in Science and Technology.

CPST. 2006b. STEM Workforce Data Project: Report No. 5, Science and Technology Salaries: Trends and Details, 1995-2005. New York: Commission on Professionals in Science and Technology.

Deloitte. 2012. The Aerospace and Defense Industry in the U.S.—A Financial and Economic Impact Study. New York: Deloitte Development LLC.

Department of Defense Office of the Under Secretary for Personnel and Readiness. 2001. Occupational Conversion Index. Arlington, Va.: U.S. Department of Defense.

Department of Homeland Security. 2007. Defense Industrial Base Critical Infrastructure and Key Resources Sector-Specific Plan as Input to the National Infrastructure Protection Plan. Washington, D.C.: Department of Homeland Security.

Endres, A. 1996. A synopsis of software engineering history: The industrial perspective. Position Papers for Dagstuhl Seminar 9635 on History of Software Engineering, edited by A. Brennecke and R. Keil-Slawik. Available at www.dagstuhl.de/Reports/96/9635.pdf.

Hedden, C.R. 2011. Aviation Week Workforce Study 2011. Arlington, Va.: Aerospace Industries Association, American Institute of Aeronautics and Astronautics, and National Defense Industries Association.

Lockard, C.B., and M. Wolf. 2012. Occupational Employment Projections to 2020. Washington, D.C.: Bureau of Labor Statistics.

Lowell, B.L., H. Salzman, H. Bernstein, and E. Henderson. 2009. Steady as She Goes? Three Generations of Students Through the Science and Engineering Pipeline. Paper presented at Annual Meetings of the Association for Public Policy Analysis and Management, Washington, D.C., October 9.

National Research Council. 2000. Forecasting Demand and Supply of Doctoral Scientists and Engineers: Report of a Workshop on Methodology. Washington, D.C.: National Academy Press.

National Research Council. 2012. Report of a Workshop on Science, Technology, Engineering, and Mathematics (STEM) Workforce Needs for the U.S. Department of Defense and the U.S. Defense Industrial Base. Washington, D.C.: The National Academies Press.

National Science Board. 2012. Science and Engineering Indicators 2012. Arlington Va.: National Science Foundation.

National Science Foundation and Division of Science Resources Statistics. 2008. Federal Scientists and Engineers: 2003-2005 (NSF 09-302). Available at http://www.nsf.gov/statistics/nsf09302/. Arlington, Va.: National Science Foundation.

Office of Personnel Management. 2009. Handbook of Occupational Groups and Families. Washington, D.C.: Office of Personnel Management.

Office of Personnel Management. 2012. Retirement Information and Services, FAQs. Avilable at http://www.opm.gov/retire/faq/pre/faq11.asp (accessed June 6, 2012).

Wyatt, I.D. 2010. Evaluating the 1996-2006 employment projections. Monthly Labor Review 133(September):33-69.

ANNEX—CHAPTER 3 TABLES

TABLE 3-1 Approaches Used to Estimate STEM Employment in Recent Reports by U.S. Government Agencies

Agency	Occupational Groups Included*	Education Coverage	Data Source(s)	Year of Estimate	STEM** Employment Estimate
Bureau of Labor Statistics, Department of Labor	• Architects • **Biological, agricultural, and environmental life scientists** • **Computer and mathematical scientists** • **Engineers** • **Physical scientists** • STEM managers • STEM postsecondary teachers • STEM sales occupations • STEM technicians	No education requirement	Bureau of Labor Statistics' Occupational Employment Statistics (OES)	2009	~8,000,000
Economics and Statistics Administration, Department of Commerce	• **Biological, agricultural, and environmental life scientists** • **Computer and mathematical scientists** • **Engineers** • **Physical scientists** • STEM managers • STEM sales occupations • STEM technicians	No education requirement	Census Bureau's American Community Survey (ACS); Bureau of Labor Statistics' Current Population Survey (CPS)	2010	~7,600,000
National Science Foundation	• **Biological, agricultural, and environmental life scientists** • **Computer and mathematical scientists** • **Engineers** • **Physical scientists** • Social scientists	Minimum of a bachelor's degree	Census Bureau's American Community Survey (ACS)	2009	4,750,000
National Science Foundation	• **Biological, agricultural, and environmental life scientists** • **Computer and mathematical scientists** • **Engineers** • **Physical scientists** • Social scientists • S&E postsecondary teachers	Minimum of a bachelor's degree	National Science Foundation's Scientists and Engineers Statistical Data System (SESTAT)	2008	4,874,000
National Science Foundation	• **Biological, agricultural, and environmental life scientists** • **Computer and mathematical scientists** • **Engineers** • **Physical scientists** • Social scientists • S&E postsecondary teachers	No education requirement	Bureau of Labor Statistics' Occupational Employment Statistics (OES)	2009	5,786,000
National Science Foundation	• **Biological, agricultural, and environmental life scientists** • **Computer and mathematical scientists** • **Engineers** • **Physical scientists** • Social scientists	No education requirement	Census Bureau's American Community Survey (ACS)	2009	6,416,000

*Occupational groups that are included in all of the STEM employment estimates are highlighted in boldface. However, the specific occupations within an occupational group that are included in a STEM estimate vary by study.

**The National Science Foundation uses the term "scientists and engineers" (S&E) for its reporting of information related to science, technology, engineering, and mathematics.

SOURCE: Compiled from information in Cover et al. (2011) and Chapter 3 of National Science Board (2012).

TABLE 3-2 Employment by STEM Occupational Group, Selected Years: 1950, 1960, 1970, 1980, 1990, and 2000 (figures in thousands)

Occupational Group	1950	1960	1970	1980	1990	2000	Annual Growth Rate, 1950-2000
All workers in U.S. reporting an occupation	60,288	69,053	81,540	105,665	124,773	138,754	1.7%
Life sciences	43	55	86	126	159	222	3.3%
Physical sciences	100	119	167	188	251	372	2.7%
Engineering	562	869	1,250	1,470	1,781	1,820	2.4%
Mathematics and information technology	26	28	377	646	1,457	3,267	10.1%
Social sciences	18	34	107	220	398	351	6.1%
Science technicians	116	255	516	190	215	299	1.9%
Engineering technicians	28	143	262	618	785	540	6.1%
ALL STEM GROUPS	894	1,503	2,764	3,459	5,046	6,871	4.2%

SOURCE: Adapted from CPST (2006a). Annual growth rates were calculated by the National Research Council using the following compound annual growth rate formula: (Ending value ÷ Beginning value)$^{1/N}$ – 1, where N is the number of periods that have elapsed between the beginning and ending values.

TABLE 3-3 Employment by STEM Occupational Group, 2010 and 2020 (projected)

		Projections, 2010-2020			
Occupational Group	Employment 2010 (000s)	Employment 2020 (000s)	Employment Growth Rate (%)	Replacement Rate (%)	Job Openings (000s)
Total, All Occupations	143,068	163,537	14.3	23.6	54,787
Life sciences	286	344	20.4	16.7	106
Physical sciences	282	318	12.7	30.6	122
Engineering	1,519	1,679	10.6	24.1	526
Mathematics and information technology*	3,542	4,320	22.0	18.6	1,437
Social sciences	306	363	18.4	32.1	155
STEM managers*	534	609	14.0	20.8	186
STEM technicians*	865	939	8.5	25.5	295
ALL STEM GROUPS*	7,333	8,571	16.9	21.7	2,827

SOURCE: Data are from the Bureau of Labor Statistics, U.S. Department of Labor, Employment Projections Program (www.bls.gov/emp/). The employment growth rate equals the number of new jobs expected over the 10-year period as a percentage of 2010 employment. The replacement rate equals the number of job openings due to replacement needs expected over the 10-year period as a percentage of 2010 employment. Projected job openings are due to the combination of growth and replacement needs. An asterisk (*) indicates that the information presented has been computed by the National Research Council.

TABLE 3-4 Aerospace Products and Parts Manufacturing (NAICS 3364) STEM Employment by Occupational Group, 2010 and 2020 (projected)

	2010			2020 (Projected)		
Occupational Group	Employment (000s)	Percent of Industry	Percent of Occupation	Employment (000s)	Percent of Industry	Percent of Occupation
Life sciences	N/A	N/A	N/A	N/A	N/A	N/A
Physical sciences	0.5	0.1	0.2	0.4	0.1	0.1
Engineering	82.2	17.2	5.4	78.4	17.0	4.7
Mathematics and information iechnology*	37.0	7.7	1.0	35.7	7.7	0.8
Social sciences	N/A	N/A	N/A	N/A	N/A	N/A
STEM managers**	9.0	1.8	1.7	8.3	1.8	1.4
STEM technicians**	15.2	3.2	1.8	14.6	3.1	1.6
ALL STEM GROUPS**	143.9	30.2	2.0	137.4	29.7	1.6
ALL OCCUPATIONS	477.1	100.0	0.3	462.4	100.0	0.3

*A small number of Mathematical Technicians may be included in these figures.

**The information presented has been computed by the National Research Council. In these cases, percent of occupation has been computed as a percent of occupational employment as provided in Table 3-3.

SOURCE: Data are from the Bureau of Labor Statistics, U.S. Department of Labor, Employment Projections Program (www.bls.gov/emp/). In some instances (indicated by "N/A") no employment counts were provided in the raw data for an entire occupational group because the group has fewer than 50 jobs, confidential data, or poor quality data.

TABLE 3-5 Aerospace Products and Parts Manufacturing (NAICS 3364) STEM Employment by Occupational Group and Occupation, 2010 and 2020 (projected)

	2010			2020 (Projected)		
Occupational Group	Employment (000s)	Percent of Industry	Percent of Occupation	Employment (000s)	Percent of Industry	Percent of Occupation
Life sciences	N/A	N/A	N/A	N/A	N/A	N/A
Physical sciences	0.5	0.1	0.2	0.4	0.1	0.1
Engineering	82.2	17.2	5.4	78.4	17.0	4.7
Aerospace Engineers	28.5	6.0	35.2	25.5	5.5	30.0
Civil Engineers	0.3	0.1	0.1	0.2	0.1	0.1
Electrical Engineers	5.7	1.2	3.7	5.7	1.2	3.5
Electronics Engineers, Except Computer	4.1	0.8	2.9	3.8	0.8	2.6
Environmental Engineers	0.3	0.1	0.6	0.3	0.1	0.4
Health and Safety Engineers, Except Mining Safety Engineers and Inspectors	0.4	0.1	1.6	0.3	0.1	1.3
Industrial Engineers	16.0	3.3	7.8	17.2	3.7	7.9
Materials Engineers	3.8	0.8	17.2	4.9	1.0	20.0
Mechanical Engineers	11.3	2.4	4.6	10.5	2.3	4.0
Engineers, All Other	11.4	2.4	7.3	9.5	2.1	5.7
Mathematics and information technology*	37.0	7.7	1.0	35.7	7.7	0.8
Computer Systems Analysts	3.5	0.7	0.6	3.2	0.7	0.5
Computer Programmers	2.0	0.4	0.5	1.6	0.4	0.4
Software Developers, Applications	6.3	1.3	1.2	5.9	1.3	0.9
Software Developers, Systems Software	15.0	3.1	3.8	15.3	3.3	2.9
Database Administrators	1.1	0.2	1.0	1.1	0.2	0.8
Network and Computer Systems Administrators	1.1	0.2	0.3	1.1	0.2	0.2
Computer Support Specialists	1.4	0.3	0.2	1.3	0.3	0.2
Information Security Analysts, Web Developers, and Computer Network Architects	2.7	0.6	0.9	2.5	0.5	0.7
Computer Occupations, All Other	2.9	0.6	1.4	2.7	0.6	1.2
Mathematical Science Occupations	1.0	0.2	0.9	0.9	0.2	0.7
Operations Research Analysts	1.0	0.2	1.5	0.9	0.2	1.2
Social sciences	N/A	N/A	N/A	N/A	N/A	N/A
STEM managers**	9.0	1.8	1.7	8.3	1.8	1.4
Computer and Information Systems Managers	2.0	0.4	0.6	1.8	0.4	0.5
Architectural and Engineering Managers	6.9	1.4	3.9	6.4	1.4	3.3
Natural Sciences Managers	0.1	0.0	0.1	0.1	0.0	0.1
STEM technicians**	15.2	3.2	1.8	14.6	3.1	1.6
Aerospace Engineering and Operations Technicians	3.0	0.6	34.4	2.5	0.5	29.1
Electrical and Electronics Engineering Technicians	2.7	0.6	1.8	2.5	0.5	1.6
Electro-Mechanical Technicians	0.3	0.1	2.1	0.3	0.1	1.9
Industrial Engineering Technicians	4.4	0.9	7.0	5.0	1.1	7.7
Mechanical Engineering Technicians	1.9	0.4	4.3	1.6	0.3	3.4
Engineering Technicians, Except Drafters, All Other	2.8	0.6	4.0	2.6	0.6	3.5
Life, Physical, and Social Science Technicians	0.1	0.0	0.0	0.1	0.0	0.0
ALL STEM GROUPS**	143.9	30.2	2.0	137.4	29.7	1.6

*A small number of Mathematical Technicians may be included in these figures.

**The information presented has been computed by the National Research Council. In these cases, percent of occupation has been computed as a percent of occupational employment as provided in Table 3-3.

SOURCE: Data are from the Bureau of Labor Statistics, U.S. Department of Labor, Employment Projections Program (www.bls.gov/emp/). Total employment for an occupational group and the sum of employment across the occupations within the occupational group may differ due to (1) rounding in the raw data and (2) the exclusion from the raw data of employment counts for occupations with fewer than 50 jobs, confidential data, or poor quality data. In some instances (indicated by "N/A") no employment counts were provided in the raw data for an entire occupational group, because the group has fewer than 50 jobs, confidential data, or poor quality data.

TABLE 3-6 Age Distribution of the Defense Industrial Base STEM Workforce

Year	<35 Years Old	35-54 Years Old	55+ Years Old	Median Age
2005	20.0%	61.4%	18.6%	45.0
2010	22.1%	55.1%	22.8%	47.0

SOURCE: Data are from the American Community Survey Public Use Microdata Sample Files, 2005 and 2010. Tabulations by the National Research Council.

TABLE 3-7 Retirements and Retirement Eligibility for the Aerospace and Defense Industry Workforce by Job Category, 2010-2014

		Percent Eligible to Retire				
Job Category	Percent Retiring, 2010	2010	2011*	2012*	2013*	2014*
Overall	1.2	13.8	14.7	17.1	13.8	22.8
Engineering	0.9	11.0	12.5	15.0	17.4	18.3
Software Dev	0.4	5.9	7.9	8.9	NA	NA
R&D	0.9	13.7	15.7	7.7	24.8	26.5
Test & Eval	0.9	12.4	14.5	16.6	NA	29.4
Enterprise IT	0.8	9.0	9.4	11.2	16.4	19.9
Eng Tech Aides	1.3	12.2	13.8	16.2	22.9	25.0
Program Mgt	1.5	13.0	14.5	16.7	25.3	22.7
Finance	0.9	9.8	10.9	12.9	19.7	20.2
Bus Dev	1.5	14.2	15.1	17.5	26.4	26.4
Supply Ch	1.1	12.1	13.8	15.4	NA	29.4
Sustainment	1.3	13.1	14.3	17.6	NA	NA
Non-Exempt	2.0	15.4	16.5	19.4	20.2	25.1

*Projected.
SOURCE: Data from Hedden (2011). Figures have been rounded to the first decimal place by the National Research Council.

TABLE 3-8 Retirements and Retirement Eligibility of the Aerospace and Defense Industry Workforce by Company Size for 2010, 2011, 2012, and 2016

	2010		2011	2012	2016
Company Size	Percent Retiring	Eligible to Retire	Projected Percent Eligible to Retire		
50,000+	1.6	14.5	17.2	20.1	32.3
10,000-49,999	1.3	14.2	16.7	19.6	33.0
1,000-9,999	1.1	16.4	16.7	19.1	29.1
Under 1,000	1.2	2.8	1.4	1.7	3.2

SOURCE: Data from Hedden (2011). Figures have been rounded to the first decimal place by the National Research Council.

TABLE 3-9 Crosswalk Between STEM Major and Minor Occupational Groups and OPM Occupational Series and Department of Defense Civilian STEM Employment by OPM Occupational Series, 2011

STEM Occupational Group	OPM Occupational Series	2011 Employment	Percent of 2011 Employment
	Computer and mathematical scientist: 52,657		
Computer/ information scientist	1550-COMPUTER SCIENCE	**5,384**	**3.6%**
	1670-EQUIPMENT SERVICES	**6,418**	**4.2%**
	2210-INFORMATION TECHNOLOGY MANAGEMENT	**35,946**	**23.8%**
Mathematical scientist	1515-OPERATIONS RESEARCH	**3,879**	**2.6%**
	1520-MATHEMATICS	836	0.6%
	1529-MATHEMATICAL STATISTICS	64	0.0%
	1530-STATISTICS	130	0.1%
	1541-CRYPTANALYSIS	0	0.0%
	Engineer: 71,123		
Aerospace engineer	0861-AEROSPACE ENGINEERING	**4,090**	**2.7%**
Chemical engineer	0893-CHEMICAL ENGINEERING	834	0.6%
Civil engineer	0810-CIVIL ENGINEERING	**8,187**	**5.4%**
Electrical/ electronics/ computer engineer	0850-ELECTRICAL ENGINEERING	**3,349**	**2.2%**
	0854-COMPUTER ENGINEERING	**3,366**	**2.2%**
	0855-ELECTRONICS ENGINEERING	**17,238**	**11.4%**
Industrial engineer	0803-SAFETY ENGINEERING	324	0.2%
	0804-FIRE PROTECTION ENGINEERING	102	0.1%
	0896-INDUSTRIAL ENGINEERING	1,167	0.8%
Mechanical engineer	0830-MECHANICAL ENGINEERING	**10,920**	**7.2%**
Other engineer	0801-GENERAL ENGINEERING	**15,470**	**10.2%**
	0806-MATERIALS ENGINEERING	819	0.5%
	0819-ENVIRONMENTAL ENGINEERING	**2,186**	**1.4%**
	0840-NUCLEAR ENGINEERING	**2,136**	**1.4%**
	0858-BIOENGINEERING & BIOMEDICAL ENGINEERING	76	0.1%
	0871-NAVAL ARCHITECTURE	815	0.5%
	0880-MINING ENGINEERING	1	0.0%
	0881-PETROLEUM ENGINEERING	1	0.0%
	0890-AGRICULTURAL ENGINEERING	6	0.0%
	0892-CERAMIC ENGINEERING*	0	0.0%
	0894-WELDING ENGINEERING*	0	0.0%
	1321-METALLURGY	36	0.0%

TABLE 3-9 Continued

STEM Occupational Group	OPM Occupational Series	2011 Employment	Percent of 2011 Employment
	Life scientist: 6,944		
Agricultural/food scientist	0028-ENVIRONMENTAL PROTECTION SPECIALIST	**2,005**	**1.3%**
	0406-AGRICULTURAL EXTENSION*	0	0.0%
	0437-HORTICULTURE	1	0.0%
	0454-RANGELAND MANAGEMENT	6	0.0%
	0457-SOIL CONSERVATION	4	0.0%
	0470-SOIL SCIENCE	4	0.0%
	0471-AGRONOMY	19	0.0%
	0487-ANIMAL SCIENCE	1	0.0%
Biological scientist	0401-GENERAL NATURAL RESOURCES MANAGEMENT AND BIOLOGICAL SCIENCES	**3,943**	**2.6%**
	0403-MICROBIOLOGY	318	0.2%
	0405-PHARMACOLOGY	29	0.0%
	0408-ECOLOGY	124	0.1%
	0410-ZOOLOGY	2	0.0%
	0413-PHYSIOLOGY	119	0.1%
	0414-ENTOMOLOGY	42	0.0%
	0415-TOXICOLOGY	50	0.0%
	0430-BOTANY	18	0.0%
	0434-PLANT PATHOLOGY	1	0.0%
	0435-PLANT PHYSIOLOGY	1	0.0%
	0440-GENETICS	1	0.0%
	0482-FISH BIOLOGY	94	0.1%
	0486-WILDLIFE BIOLOGY	162	0.1%
	Physical scientist: 7,381		
Chemist, except biochemist	1320-CHEMISTRY	**1,580**	**1.0%**
Earth/atmospheric/ocean scientist	1310-PHYSICS	**1,680**	**1.1%**
	1313-GEOPHYSICS	80	0.1%
	1315-HYDROLOGY	68	0.0%
	1330-ASTRONOMY AND SPACE SCIENCE	96	0.1%
	1340-METEOROLOGY	318	0.2%
	1350-GEOLOGY	367	0.2%
	1360-OCEANOGRAPHY	279	0.2%
	1372-GEODESY	10	0.0%

continued

TABLE 3-9 Continued

STEM Occupational Group	OPM Occupational Series	2011 Employment	Percent of 2011 Employment
Other physical scientist	1301-GENERAL PHYSICAL SCIENCE	**2,410**	**1.6%**
	1306-HEALTH PHYSICS	493	0.3%
	Social scientist: 13,172		
Economics/business	0110-ECONOMIST	247	0.2%
	0135-FOREIGN AGRICULTURAL AFFAIRS	0	0.0%
	1140-TRADE SPECIALIST	0	0.0%
	1146-AGRICULTURAL MARKETING	0	0.0%
	1147-AGRICULTURAL MARKET REPORTING	0	0.0%
	2110-TRANSPORTATION INDUSTRY ANALYSIS	1	0.0%
Political scientist	0130-FOREIGN AFFAIRS	518	0.3%
	0131-INTERNATIONAL RELATIONS	157	0.1%
Psychologist	0180-PSYCHOLOGY	**1,589**	**1.1%**
Sociologist/ anthropologist	0184-SOCIOLOGY	12	0.0%
	0190-GENERAL ANTHROPOLOGY	40	0.0%
Other social scientist	0101-SOCIAL SCIENCE	**3,503**	**2.3%**
	0106-UNEMPLOYMENT INSURANCE	0	0.0%
	0132-INTELLIGENCE	**6,619**	**4.4%**
	0136-INTERNATIONAL COOPERATION	1	0.0%
	0140-WORKFORCE RESEARCH AND ANALYSIS	0	0.0%
	0150-GEOGRAPHY	228	0.2%
	0160-CIVIL RIGHTS ANALYSIS	0	0.0%
	0193-ARCHEOLOGY	242	0.2%
	1730-EDUCATION RESEARCH	15	0.0%
	Total DOD civilian STEM employment: 151,277		

NOTE: Figures are as of the fiscal year-end (September 30, 2011). OPM occupations with 1 percent or more of DOD STEM employment are highlighted in boldface.

*Series has been cancelled as of the 2011 fiscal year.

SOURCE: Data provided by the Defense Manpower Data Center. Tabulations by the National Research Council.

TABLE 3-10 Department of Defense Civilian STEM Employment by Major Occupational Group, 2001-2011

STEM Occupational Group	2001	2002	2003	2004	2005	2006	2007	2008	2009	2010	2011	Annual Growth Rate, 2001-2011
Computer and mathematical scientist	37,604	40,624	41,025	41,284	41,968	41,915	41,202	42,771	46,082	50,693	52,657	3.4%
Engineer	57,449	59,296	60,586	60,970	60,643	60,433	60,182	61,386	66,104	70,462	71,123	2.2%
Life scientist	4,847	5,033	5,160	5,271	5,411	5,538	5,752	6,208	6,742	6,979	6,944	3.7%
Physical scientist	6,729	6,771	6,803	6,806	6,686	6,615	6,543	6,715	7,141	7,452	7,381	0.9%
Social scientist	6,996	7,279	7,715	7,924	8,454	8,822	9,104	9,759	10,447	11,533	13,172	6.5%
Total DOD civilian STEM employment	113,625	119,003	121,289	122,255	123,162	123,323	122,783	126,839	136,516	147,119	151,277	2.9%

NOTE: Figures are as of the fiscal year-end (e.g., fiscal year 2011 is as of September 30, 2011). Growth rates are calculated using the compound annual growth rate formula: (Ending value ÷ Beginning value)$^{1/N}$ – 1, where N is the number of periods that have elapsed between the beginning and ending values.

SOURCE: Data provided by the Defense Manpower Data Center. Tabulations by the National Research Council.

TABLE 3-11 Department of Defense 20 Largest Civilian STEM Occupations, 2011

OPM Occupational Series	2011 Employment	Percent of 2011 Employment
2210-INFORMATION TECHNOLOGY MANAGEMENT	35,946	23.8
0855-ELECTRONICS ENGINEERING	17,238	11.4
0801-GENERAL ENGINEERING	15,470	10.2
0830-MECHANICAL ENGINEERING	10,920	7.2
0810-CIVIL ENGINEERING	8,187	5.4
0132-INTELLIGENCE	6,619	4.4
1670-EQUIPMENT SERVICES	6,418	4.2
1550-COMPUTER SCIENCE	5,384	3.6
0861-AEROSPACE ENGINEERING	4,090	2.7
0401-GENERAL NATURAL RESOURCES MANAGEMENT AND BIOLOGICAL SCIENCES	3,943	2.6
1515-OPERATIONS RESEARCH	3,879	2.6
0101-SOCIAL SCIENCE	3,503	2.3
0854-COMPUTER ENGINEERING	3,366	2.2
0850-ELECTRICAL ENGINEERING	3,349	2.2
1301-GENERAL PHYSICAL SCIENCE	2,410	1.6
0819-ENVIRONMENTAL ENGINEERING	2,186	1.4
0840-NUCLEAR ENGINEERING	2,136	1.4
0028-ENVIRONMENTAL PROTECTION SPECIALIST	2,005	1.3
1310-PHYSICS	1,680	1.1
0180-PSYCHOLOGY	1,589	1.1

NOTE: Figures are as of the fiscal year-end (September 30, 2011).
SOURCE: Data provided by the Defense Manpower Data Center. Tabulations by the National Research Council.

TABLE 3-12 Department of Defense 20 Fastest-Growing Civilian STEM Occupations, 2001-2011

OPM Occupational Series	Annual Growth Rate, 2001-2011
0130-FOREIGN AFFAIRS	11.3%
0150-GEOGRAPHY	10.6%
0401-GENERAL NATURAL RESOURCES MANAGEMENT AND BIOLOGICAL SCIENCES	7.9%
0132-INTELLIGENCE	7.6%
0858-BIOENGINEERING & BIOMEDICAL ENGINEERING	6.6%
1550-COMPUTER SCIENCE	6.0%
0180-PSYCHOLOGY	5.6%
0854-COMPUTER ENGINEERING	5.6%
0131-INTERNATIONAL RELATIONS	5.4%
0101-SOCIAL SCIENCE	5.3%
0804-FIRE PROTECTION ENGINEERING	4.8%
0850-ELECTRICAL ENGINEERING	4.7%
0801-GENERAL ENGINEERING	4.3%
0193-ARCHEOLOGY	4.1%
1306-HEALTH PHYSICS	4.1%
2210-INFORMATION TECHNOLOGY MANAGEMENT	3.7%
1515-OPERATIONS RESEARCH	3.5%
0840-NUCLEAR ENGINEERING	3.5%
0482-FISH BIOLOGY	3.4%
0830-MECHANICAL ENGINEERING	3.0%

NOTE: Figures are as of the fiscal year-end (September 30, 2011). The table excludes OPM occupations with less than 0.05% of 2011 Department of Defense civilian STEM employment. Growth rates are calculated using the compound annual growth rate formula: (Ending value ÷ Beginning value)$^{1/N}$ – 1, where N is the number of periods that have elapsed between the beginning and ending values.

SOURCE: Data provided by the Defense Manpower Data Center. Tabulations by the National Research Council.

TABLE 3-13 Department of Defense Civilian STEM Employment by OPM Occupational Series, 2001-2011

STEM Occupational Group	OPM Occupational Series	2001	2002	2003	2004	2005	2006	2007	2008	2009	2010	2011	Annual Growth Rate, 2001-2011
Computer and mathematical scientist													
Computer/ information scientist	1550-COMPUTER SCIENCE	3,012	3,442	3,734	3,907	4,025	4,150	4,266	4,452	4,815	5,176	5,384	6.0%
	1670-EQUIPMENT SERVICES	5,771	5,791	5,612	5,473	5,725	5,656	5,701	5,914	6,212	6,427	6,418	1.1%
	2210-INFORMATION TECHNOLOGY MANAGEMENT	24,897	27,487	27,745	27,952	28,261	28,162	27,247	28,225	30,567	34,283	35,946	3.7%
Mathematical scientist	1515-OPERATIONS RESEARCH	2,745	2,756	2,813	2,868	2,927	2,966	3,020	3,200	3,465	3,770	3,879	3.5%
	1520-MATHEMATICS	1,010	981	955	917	876	826	807	811	834	835	836	-1.9%
	1529-MATHEMATICAL STATISTICS	60	56	56	53	50	49	47	50	50	59	64	0.6%
	1530-STATISTICS	109	111	110	114	104	106	114	119	139	143	130	1.8%
	1541-CRYPTANALYSIS	0	0	0	0	0	0	0	0	0	0	0	0.0%
Engineer													
Aerospace engineer	0861-AEROSPACE ENGINEERING	3,421	3,536	3,631	3,621	3,559	3,520	3,543	3,644	3,865	4,052	4,090	1.8%
Chemical engineer	0893-CHEMICAL ENGINEERING	674	731	770	788	785	769	765	803	843	845	834	2.2%
Civil engineer	0810-CIVIL ENGINEERING	6,836	6,914	6,984	6,787	6,522	6,384	6,314	6,554	7,621	8,174	8,187	1.8%
Electrical/ electronics/ computer engineer	0850-ELECTRICAL ENGINEERING	2,113	2,276	2,405	2,447	2,441	2,478	2,514	2,633	2,949	3,266	3,349	4.7%
	0854-COMPUTER ENGINEERING	1,961	2,285	2,518	2,804	2,916	2,919	2,981	3,059	3,212	3,361	3,366	5.6%
	0855-ELECTRONICS ENGINEERING	17,060	17,293	17,484	17,386	17,167	16,994	16,687	16,614	17,011	17,513	17,238	0.1%
Industrial engineer	0803-SAFETY ENGINEERING	284	277	299	294	293	303	295	300	303	324	324	1.3%
	0804-FIRE PROTECTION ENGINEERING	64	76	87	81	75	73	72	68	83	96	102	4.8%
	0896-INDUSTRIAL ENGINEERING	1,053	1,072	1,043	1,016	1,012	1,004	984	959	1,045	1,173	1,167	1.0%

continued

TABLE 3-13 Continued

STEM Occupational Group	OPM Occupational Series	2001	2002	2003	2004	2005	2006	2007	2008	2009	2010	2011	Annual Growth Rate, 2001-2011
Mechanical engineer	0830-MECHANICAL ENGINEERING	8,134	8,599	8,943	9,097	9,031	9,069	9,045	9,260	10,192	10,902	10,920	3.0%
Other engineer	0801-GENERAL ENGINEERING	10,133	10,324	10,456	10,662	10,924	11,158	11,428	11,963	13,209	14,776	15,470	4.3%
	0806-MATERIALS ENGINEERING	732	762	783	792	792	759	751	757	783	788	819	1.1%
	0819-ENVIRONMENTAL ENGINEERING	2,582	2,588	2,587	2,542	2,498	2,366	2,239	2,111	2,154	2,212	2,186	-1.7%
	0840-NUCLEAR ENGINEERING	1,514	1,656	1,708	1,773	1,777	1,778	1,713	1,791	1,921	2,034	2,136	3.5%
	0858-BIOENGINEERING & BIOMEDICAL ENGINEERING	40	45	46	52	54	53	56	63	66	71	76	6.6%
	0871-NAVAL ARCHITECTURE	725	744	724	711	688	698	699	710	758	813	815	1.2%
	0880-MINING ENGINEERING	0	0	0	0	0	0	0	0	0	0	1	—
	0881-PETROLEUM ENGINEERING	4	2	2	1	1	1	.	1	1	1	1	-12.9%
	0890-AGRICULTURAL ENGINEERING	2	2	3	2	2	3	3	3	4	4	6	11.6%
	0892-CERAMIC ENGINEERING*	13	14	13	14	14	12	11	11	9	5	0	—
	0894-WELDING ENGINEERING*	39	38	39	45	43	44	42	46	37	15	0	—
	1321-METALLURGY	65	62	61	55	49	48	40	36	38	37	36	-5.7%
					Life scientist								
Agricultural/ food scientist	0028-ENVIRONMENTAL PROTECTION SPECIALIST	2,120	2,136	2,149	2,110	2,063	2,001	1,934	1,966	2,038	2,094	2,005	-0.6%
	0406-AGRICULTURAL EXTENSION*	0	0	0	0	0	0	0	0	0	0	0	—
	0437-HORTICULTURE	3	3	3	2	2	2	3	3	3	3	1	-10.4%
	0454-RANGELAND MANAGEMENT	11	6	7	7	7	4	5	5	7	6	6	-5.9%
	0457-SOIL CONSERVATION	9	10	10	10	10	8	7	8	5	4	4	-7.8%

	0470-SOIL SCIENCE	6	7	8	8	6	5	4	5	4	4	4	-4.0%
	0471-AGRONOMY	31	28	28	26	25	21	21	20	20	18	19	-4.8%
	0487-ANIMAL SCIENCE	1	1	1	1	2	2	2	1	1	1	1	0.0%
Biological scientist	0401-GENERAL NATURAL RESOURCES MANAGEMENT AND BIOLOGICAL SCIENCES	1,843	1,992	2,107	2,226	2,402	2,615	2,908	3,313	3,727	3,889	3,943	7.9%
	0403-MICROBIOLOGY	258	275	266	287	297	300	297	303	323	316	318	2.1%
	0405-PHARMACOLOGY	25	26	21	23	25	26	25	25	26	29	29	1.5%
	0408-ECOLOGY	126	115	121	124	122	117	111	113	125	124	124	-0.2%
	0410-ZOOLOGY	5	6	5	5	5	4	3	3	2	2	2	-8.8%
	0413-PHYSIOLOGY	106	109	106	114	114	117	112	117	120	120	119	1.2%
	0414-ENTOMOLOGY	48	51	56	51	48	44	43	37	35	41	42	-1.3%
	0415-TOXICOLOGY	38	42	45	50	47	48	52	52	51	54	50	2.8%
	0430-BOTANY	20	20	20	18	18	18	17	17	20	20	18	-1.0%
	0434-PLANT PATHOLOGY	1	1	1	1	1	1	1	1	1	1	1	0.0%
	0435-PLANT PHYSIOLOGY	1	1	1	1	1	1	1	1	1	1	1	0.0%
	0440-GENETICS	2	2	2	1	0	0	0	0	0	1	1	-6.7%
	0482-FISH BIOLOGY	67	70	72	73	77	73	70	74	84	89	94	3.4%
	0486-WILDLIFE BIOLOGY	126	132	131	133	139	131	136	144	149	162	162	2.5%
					Physical scientist								
Chemist, except biochemist	1320-CHEMISTRY	1,529	1,514	1,502	1,511	1,499	1,475	1,467	1,510	1,553	1,586	1,580	0.3%
Earth/ atmospheric/ ocean scientist	1310-PHYSICS	1,824	1,841	1,811	1,768	1,711	1,671	1,661	1,647	1,700	1,751	1,680	-0.8%
	1313-GEOPHYSICS	101	105	102	98	97	97	91	86	85	83	80	-2.3%
	1315-HYDROLOGY	74	77	75	82	77	74	70	67	72	72	68	-0.8%
	1330-ASTRONOMY AND SPACE SCIENCE	117	115	109	107	101	100	99	93	95	97	96	-2.0%
	1340-METEOROLOGY	280	273	281	284	284	286	284	288	293	314	318	1.3%
	1350-GEOLOGY	316	323	329	320	304	293	288	309	352	360	367	1.5%
	1360-OCEANOGRAPHY	322	317	317	301	302	288	277	283	288	300	279	-1.4%
	1372-GEODESY	7	7	7	7	7	6	7	8	8	8	10	3.6%

continued

TABLE 3-13 Continued

STEM Occupational Group	OPM Occupational Series	2001	2002	2003	2004	2005	2006	2007	2008	2009	2010	2011	Annual Growth Rate, 2001-2011
Other physical scientist	1301-GENERAL PHYSICAL SCIENCE	1,828	1,870	1,923	1,987	1,950	1,956	1,921	2,015	2,255	2,406	2,410	2.8%
	1306-HEALTH PHYSICS	331	329	347	341	354	369	378	409	440	475	493	4.1%
Social scientist													
Economics/ business	0110-ECONOMIST	252	255	254	249	240	226	210	225	236	241	247	-0.2%
	0135-FOREIGN AGRICULTURAL AFFAIRS	0	0	0	0	0	0	0	0	0	0	0	—
	1140-TRADE SPECIALIST	2	2	1	1	1	0	0	0	0	0	0	—
	1146-AGRICULTURAL MARKETING	0	0	0	0	0	0	0	0	0	0	0	—
	1147-AGRICULTURAL MARKET REPORTING	0	0	0	0	0	0	0	0	0	0	0	—
	2110-TRANSPORTATION INDUSTRY ANALYSIS	0	3	2	0	0	0	0	0	0	1	1	—
Political scientist	0130-FOREIGN AFFAIRS	178	195	195	208	215	241	280	329	390	439	518	11.3%
	0131-INTERNATIONAL RELATIONS	93	100	109	108	112	114	112	129	135	148	157	5.4%
Psychologist	0180-PSYCHOLOGY	923	917	915	938	968	1,011	1,006	1,062	1,201	1,425	1,589	5.6%
Sociologist/ anthropologist	0184-SOCIOLOGY	6	7	8	7	4	6	5	5	9	14	12	7.2%
	0190-GENERAL ANTHROPOLOGY	30	29	33	34	37	40	39	37	37	43	40	2.9%
Other social scientist	0101-SOCIAL SCIENCE	2,085	2,146	2,198	2,176	2,376	2,446	2,389	2,693	2,879	3,286	3,503	5.3%
	0106-UNEMPLOYMENT INSURANCE											—	—
	0132-INTELLIGENCE	3,181	3,363	3,712	3,900	4,191	4,417	4,735	4,940	5,148	5,477	6,619	7.6%
	0136-INTERNATIONAL COOPERATION	0	0	0	0	1	0	0	2	1	1	1	—

0140-WORKFORCE RESEARCH AND ANALYSIS	0	0	0	0	0	0	0	0	0	0	0	—
0150-GEOGRAPHY	83	94	109	120	126	141	145	159	194	214	228	10.6%
0160-CIVIL RIGHTS ANALYSIS	1	1	1	0	0	0	0	0	0	0	0	—
0193-ARCHEOLOGY	162	167	178	183	183	180	182	178	216	234	242	4.1%
1730-EDUCATION RESEARCH	0	0	0	0	0	0	1	0	1	10	15	—
Total DOD civilian STEM employment	113,625	119,003	121,289	122,255	123,162	123,323	122,783	126,839	136,516	147,119	151,277	2.9%

NOTE: Figures are as of the fiscal year-end (e.g., fiscal year 2011 is as of September 30, 2011). Growth rates are calculated using the compound annual growth rate formula: (Ending value ÷ Beginning value)$^{1/N}$ – 1, where N is the number of periods that have elapsed between the beginning and ending values.

*Series has been cancelled as of the 2011 fiscal year.

SOURCE: Data provided by the Defense Manpower Data Center. Tabulations by the National Research Council.

TABLE 3-14 Department of Defense Civilian STEM Employment by Department and Major Occupational Group, 2011

	Department of the Air Force		Department of the Army		Department of the Navy		Department of Defense Agencies	
STEM Occupational Group	2011 Empl	Percent of 2011 Empl	2011 Empl	Percent of 2011 Empl	2011 Empl	Percent of 2011 Empl	2011 Empl	Percent of 2011 Empl
Computer and mathematical scientist	11,409	38.4%	18,136	32.5%	15,287	28.9%	7,825	60.9%
Engineer	13,216	44.4%	23,795	42.7%	30,763	58.1%	3,349	26.1%
Life scientist	506	1.7%	4,824	8.7%	1,209	2.3%	405	3.2%
Physical scientist	1,173	3.9%	2,783	5.0%	3,172	6.0%	253	2.0%
Social scientist	3,433	11.5%	6,222	11.2%	2,494	4.7%	1,023	8.0%
Total DOD civilian STEM employment	29,737	100.0%	55,760	100.0%	52,925	100.0%	12,855	100.0%

NOTE: Figures are as of the fiscal year-end (September 30, 2011).
SOURCE: Data provided by the Defense Manpower Data Center. Tabulations by the National Research Council.

TABLE 3-15 Crosswalk Between Highest Degree Attained and OPM's Classification of Educational Attainment

Highest Degree Attained	OPM's Classification of Educational Attainment
Degree unknown	Invalid
Less than bachelor's degree	No formal education or did not complete elementary school
	Elementary school completed–no high school
	Some high school
	High School or certificate of equivalency
	Terminal occupational program–did not complete
	Terminal occupational program
	Less than one year college
	One year college
	Two years college
	Associate Degree
	Three year college
	Four years college
Bachelor's degree	Bachelor's degree
	Post-Bachelor's
Professional degree	First professional
	Post-first professional degree
Master's degree	Master's degree
	Post-Master's
	Sixth-year degree
	Post-sixth year
Doctoral degree	Doctoral degree
	Post-Doctorate

TABLE 3-16 Most Common Fields of Study (4-digit CIP code) for Department of Defense Civilian STEM Workforce with a Postsecondary Degree, 2011

Field of Study (4-digit CIP code and CIP title)		2011 Employment	Percent of 2011 Employment
14.10	Electrical, Electronics and Communications Engineering	17,202	14.1
14.19	Mechanical Engineering	13,852	11.4
14.08	Civil Engineering	8,290	6.8
52.02	Business Administration, Management and Operations	7,206	5.9
11.01	Computer and Information Sciences, General	5,400	4.4
11.07	Computer Science	4,631	3.8
14.02	Aerospace, Aeronautical and Astronautical Engineering	3,543	2.9
14.01	Engineering, General	3,125	2.6
40.08	Physics	2,291	1.9
14.09	Computer Engineering, General	2,184	1.8
14.35	Industrial Engineering	2,171	1.8
15.03	Electrical Engineering Technologies/Technicians	2,110	1.7
27.01	Mathematics	1,947	1.6
14.07	Chemical Engineering	1,942	1.6
24.01	Liberal Arts and Sciences, General Studies, and Humanities	1,712	1.4
40.05	Chemistry	1,711	1.4
26.01	Biology, General	1,521	1.2
14.99	Engineering, Other	1,370	1.1
11.04	Information Science/Studies	1,301	1.1
40.06	Geological and Earth Sciences/Geosciences	1,208	1.0
14.27	Systems Engineering	1,164	1.0

NOTE: Figures are as of the fiscal year-end (September 30, 2011). The table includes those with a highest degree of associate's degree or higher. Instructional programs are based on an individual's highest educational attainment from an accredited institution.

SOURCE: Data provided by the Defense Manpower Data Center. Tabulations by the National Research Council.

TABLE 3-17 Field of Study (2-digit CIP code) for Department of Defense Civilian STEM Workforce with a Postsecondary Degree, 2011

Field of study (2-digit CIP code and CIP title)		2011 Employment	Percent of 2011 Employment
Unknown	Unknown	431	0.4
01	Agriculture, Agriculture Operations, & Related Sciences	330	0.3
03	Natural Resources And Conservation	1,792	1.5
04	Architecture And Related Services	391	0.3
05	Area, Ethnic, Cultural, And Gender Studies	191	0.2
09	Communication, Journalism, And Related Programs	380	0.3
10	Communications Technologies/Technicians And Support Services	116	0.1
11	Computer And Information Sciences And Support Services	13,444	11.0
12	Personal And Culinary Services	9	0.0
13	Education	1,494	1.2
14	Engineering	60,435	49.5
15	Engineering Technologies/Technicians	5,310	4.4
16	Foreign Languages, Literatures, And Linguistics	172	0.1
19	Family And Consumer Sciences/Human Sciences	112	0.1
22	Legal Professions And Studies	217	0.2
23	English Language And Literature/Letters.	167	0.1
24	Liberal Arts And Sciences, General Studies, And Humanities	1,712	1.4
25	Library Science	28	0.0
26	Biological And Biomedical Sciences	3,286	2.7
27	Mathematics And Statistics	2,601	2.1
29	Military Technologies	337	0.3
30	Multi/Interdisciplinary Studies	971	0.8
31	Parks, Recreation, Leisure And Fitness Studies	308	0.3
38	Philosophy And Religious Studies	298	0.2
39	Theology And Religious Vocations	111	0.1
40	Physical Sciences	6,242	5.1
41	Science Technologies/Technicians	193	0.2
42	Psychology	2,188	1.8
43	Security And Protective Services	567	0.5
44	Public Administration And Social Service Professions	1,198	1.0
45	Social Sciences	3,104	2.5
46	Construction Trades	42	0.0
47	Mechanic And Repair Technologies/Technicians	349	0.3
48	Precision Production	12	0.0
49	Transportation And Materials Moving	352	0.3
50	Visual And Performing Arts	304	0.2
51	Health Professions And Related Clinical Sciences	674	0.6
52	Business, Management, Marketing, & Related Support Services	11,754	9.6
54	History	389	0.3
60	Residency Programs	7	0.0

NOTE: Figures are as of the fiscal year-end (September 30, 2011). The 2-digit CIP represents the most general groupings of related programs. The table includes those with a highest degree of associate's degree or higher. Instructional programs are based on an individual's highest educational attainment from an accredited institution.

SOURCE: Data provided by the Defense Manpower Data Center. Tabulations by the National Research Council.

TABLE 3-18 Field of Study (2-digit CIP code) by Major Occupational Group for Department of Defense Civilian STEM Workforce with a Postsecondary Degree, 2011

Field of Study (2-digit CIP code and CIP title)		Computer and Mathematical Scientist	Engineer	Life Scientist	Physical Scientist	Social Scientist
Unknown	Unknown	0.3%	0.1%	1.4%	0.4%	1.7%
01	Agriculture, Agriculture Operations, & Related Sciences	0.1%	0.0%	3.4%	0.8%	0.2%
03	Natural Resources And Conservation	0.2%	0.2%	22.2%	4.3%	0.3%
04	Architecture And Related Services	0.1%	0.4%	0.6%	0.2%	0.2%
05	Area, Ethnic, Cultural, And Gender Studies	0.1%	0.0%	0.1%	0.0%	1.7%
09	Communication, Journalism, And Related Programs	0.8%	0.1%	0.2%	0.0%	1.0%
10	Communications Technologies/Technicians And Support Services	0.2%	0.0%	0.1%	0.0%	0.3%
11	Computer And Information Sciences And Support Services	38.8%	1.9%	0.5%	0.9%	2.4%
12	Personal And Culinary Services	0.0%	0.0%	0.0%	0.0%	0.0%
13	Education	2.3%	0.2%	1.3%	0.7%	5.7%
14	Engineering	8.4%	81.7%	2.8%	10.1%	2.0%
15	Engineering Technologies/Technicians	3.4%	5.8%	1.0%	1.3%	0.8%
16	Foreign Languages, Literatures, And Linguistics	0.2%	0.0%	0.1%	0.0%	1.1%
19	Family And Consumer Sciences/Human Sciences	0.1%	0.0%	0.1%	0.1%	0.5%
22	Legal Professions And Studies	0.2%	0.1%	0.5%	0.1%	0.8%
23	English Language And Literature/Letters.	0.3%	0.0%	0.1%	0.0%	0.4%
24	Liberal Arts And Sciences, General Studies, And Humanities	3.4%	0.2%	1.8%	0.6%	4.1%
25	Library Science	0.1%	0.0%	0.0%	0.0%	0.1%
26	Biological And Biomedical Sciences	0.8%	0.2%	39.8%	8.2%	0.8%
27	Mathematics And Statistics	6.9%	0.5%	0.1%	1.2%	0.5%
29	Military Technologies	0.3%	0.1%	0.0%	0.2%	1.7%
30	Multi/Interdisciplinary Studies	1.0%	0.4%	1.7%	1.2%	1.9%
31	Parks, Recreation, Leisure And Fitness Studies	0.1%	0.0%	4.8%	0.0%	0.1%
38	Philosophy And Religious Studies	0.2%	0.1%	0.6%	0.6%	1.2%
39	Theology And Religious Vocations	0.2%	0.0%	0.1%	0.0%	0.4%
40	Physical Sciences	1.4%	1.8%	3.4%	60.8%	0.9%
41	Science Technologies/Technicians	0.2%	0.1%	0.3%	0.6%	0.2%
42	Psychology	1.0%	0.0%	0.9%	0.2%	19.3%
43	Security And Protective Services	0.6%	0.1%	0.7%	0.2%	2.8%
44	Public Administration And Social Service Professions	0.8%	0.3%	1.3%	0.2%	6.6%
45	Social Sciences	2.5%	0.1%	3.0%	2.0%	20.8%
46	Construction Trades	0.0%	0.0%	0.0%	0.0%	0.0%
47	Mechanic And Repair Technologies/Technicians	1.0%	0.0%	0.1%	0.0%	0.1%
48	Precision Production	0.0%	0.0%	0.0%	0.0%	0.0%
49	Transportation And Materials Moving	0.5%	0.2%	0.6%	0.5%	0.3%
50	Visual And Performing Arts	0.7%	0.0%	0.3%	0.0%	0.6%
51	Health Professions And Related Clinical Sciences	0.4%	0.1%	2.1%	1.6%	2.8%
52	Business, Management, Marketing, & Related Support Services	21.9%	4.9%	3.9%	2.7%	13.3%
54	History	0.4%	0.0%	0.3%	0.1%	2.3%
60	Residency Programs	0.0%	0.0%	0.1%	0.0%	0.0%

NOTE: Figures are as of the fiscal year-end (September 30, 2011). Fields of study that represent 3 percent or more of the degrees for a given occupational group are highlighted. The table includes those with a highest degree of associate's degree or higher. Instructional programs are based on an individual's highest educational attainment from an accredited institution.

SOURCE: Data provided by the Defense Manpower Data Center. Tabulations by the National Research Council.

TABLE 3-19 Age Distribution of Department of Defense Civilian STEM Workforce, 2001-2011

Year	22-25	26-30	31-35	36-40	41-45	46-50	51-55	56-60	61-65	66+	Median Age	Average Age
2001	2.3%	3.9%	7.8%	17.0%	19.0%	17.1%	18.0%	9.8%	3.8%	1.3%	45	45.6
2002	3.4%	4.4%	7.2%	15.1%	19.5%	17.4%	17.5%	10.3%	3.9%	1.4%	46	45.5
2003	4.1%	5.0%	6.9%	13.5%	19.8%	17.8%	16.3%	11.1%	4.1%	1.5%	46	45.4
2004	4.3%	5.8%	6.7%	11.9%	20.0%	18.2%	16.1%	11.2%	4.3%	1.5%	46	45.5
2005	3.9%	6.7%	6.6%	10.6%	19.6%	19.1%	16.0%	11.5%	4.4%	1.6%	46	45.6
2006	3.5%	7.6%	6.6%	9.5%	18.6%	19.9%	16.2%	11.7%	4.6%	1.7%	46	45.7
2007	3.4%	8.2%	6.8%	9.0%	17.0%	20.8%	16.7%	11.5%	4.9%	1.8%	47	45.9
2008	3.8%	8.7%	7.4%	8.6%	15.4%	20.8%	16.9%	11.0%	5.3%	1.9%	47	45.7
2009	4.7%	9.3%	8.4%	8.5%	13.6%	20.2%	16.8%	10.9%	5.5%	2.1%	47	45.4
2010	4.9%	9.6%	9.4%	8.7%	12.2%	19.5%	17.1%	10.7%	5.6%	2.2%	47	45.2
2011	4.2%	9.6%	10.6%	8.9%	11.5%	18.7%	17.7%	10.9%	5.6%	2.2%	47	45.3

NOTE: Figures are as of the fiscal year-end (September 30, 2011). Age category percentages exclude employees under the age of 22 and those whose age is unknown.

SOURCE: Data provided by the Defense Manpower Data Center. Tabulations by the National Research Council.

TABLE 3-20 Age Distribution of Department of Defense Civilian STEM Workforce by Major Occupational Group, 2011

Year	22-25	26-30	31-35	36-40	41-45	46-50	51-55	56-60	61-65	66+	Median Age	Average Age
Computer and mathematical scientist	2.2%	6.9%	9.6%	9.7%	13.1%	20.0%	19.3%	12.1%	5.5%	1.6%	48	46.3
Engineer	6.6%	12.3%	11.1%	7.6%	10.0%	18.8%	17.1%	9.3%	4.8%	2.3%	46	44.0
Life scientist	2.3%	7.4%	11.3%	11.6%	12.8%	14.9%	16.1%	14.0%	7.7%	1.9%	47	46.2
Physical scientist	2.3%	7.3%	9.6%	9.4%	10.2%	15.9%	18.1%	13.9%	8.1%	5.1%	49	47.7
Social scientist	1.7%	8.4%	12.0%	11.0%	13.5%	16.2%	15.4%	11.9%	7.2%	2.6%	47	45.9
Total DOD civilian STEM employment	4.2%	9.6%	10.6%	8.9%	11.5%	18.7%	17.7%	10.9%	5.6%	2.2%	47	45.3

NOTE: Figures are as of the fiscal year-end (September 30, 2011). Age category percentages exclude employees under the age of 22 and those whose age is unknown.

SOURCE: Data provided by the Defense Manpower Data Center. Tabulations by the National Research Council.

TABLE 3-21 Department of Defense Occupation Codes Identified as STEM by DMDC

DOD Classification Code	DOD Classification Code Description
123200	Analysis
169000	Other Mechanical and Electrical Equipment, General
230100	Intelligence, General
240100	Construction and Utilities
240200	Electrical/Electronic
240400	Aviation Maintenance and Allied
240700	Ship Construction and Maintenance
241000	Safety
241100	Chemical
241300	Surveying and Mapping
241400	Engineering and Maintenance Officers, Other
250100	Physical Scientists
250200	Meteorologists
250400	Social Scientists
251000	Mathematicians and Statisticians
251100	Educators and Instructors
260800	Biomedical Sciences and Allied Health Officers
260802	Biomedical Laboratory Services
260803	Environmental Health Services
260805	Pharmacy
260807	Psychology and Social Work
260814	Biochemistry
260829	Psychology, Clinical
270100	Administrators, General
270300	Manpower and Personnel
270500	Data Processing
280200	Supply
280300	Transportation

4

Limitations to Meeting the Workforce Needs of DOD and the Industrial Base

The limitations faced by the U.S. Department of Defense (DOD) and its industrial base in meeting their science, technology, engineering, and mathematics (STEM) workforce needs in both the near and long term are discussed in this chapter. While there is no evidence that a shortage of workers with the STEM skills necessary to meet the workforce needs of DOD and the industrial base *currently* exists, except in selected areas such as cybersecurity and selected intelligence fields, meeting the workforce needs associated with emerging technologies in the light of existing workforce trends and DOD policies could be problematic. First this chapter examines some of the supply and demand issues shaping the limitations likely to be faced by DOD and the industrial base in the coming years, and it then recommends some approaches that DOD might take to mitigate these limitations.[1]

SUPPLY-SIDE ISSUES

One overarching issue is whether high-performing students enter the STEM pipeline in sufficient numbers to meet the growing demand for STEM-educated workers, as discussed in Chapter 3. Data from the Programme for International Student Assessment (PISA) (OCED, 2011)[2] on the science and mathematics literacy of 15-year-olds worldwide suggest that while the proficiency in math, science, and reading of U.S. students lies in the middle rank of member countries of the Organisation for Economic Cooperation and Development (OECD), the percentage of top-performing students achieving level 5 or 6 (the two highest) is nonetheless high compared to that of other member countries; moreover, the United States produces twice as many high-performing students in absolute terms as does the next largest producer—Japan, with 40 percent of the U.S. population (Salzman, 2012; Salzman and Lowell, 2008). Data on the supply side of the pipeline show the following:

- Over the period 1993-2009, interest in pursuing a STEM degree in college remained relatively stable, with the percent of college freshmen intending to major in a STEM field ranging from 21 to 26 percent. 2010 saw a

[1]In this chapter, unless otherwise noted, STEM includes the physical sciences, biological/agricultural sciences, mathematics/statistics, computer sciences, and engineering.

[2]Further information is available at http://www.pisa.oecd.org.

new high of 27 percent, largely due to increases in those intending to study the biological/agricultural sciences and engineering.[3]

- With the exception of computer science, interest in specific STEM bachelor's degrees has increased consistently over time, with biological/agricultural sciences and engineering the most popular. In computer science, the number of bachelor's degrees awarded increased dramatically from 1998 to 2004 but fell sharply through 2008 and remained flat in 2009.[4]
- The percent of all bachelor's degrees awarded in a STEM field has been stable, ranging from 16 percent to 17 percent over the period 2000-2009.[5]
- Comparing STEM degrees awarded, on the one hand, to freshmen intentions, on the other, suggests that many students who enter college intending to get a STEM degree do not ultimately graduate with one. For example, approximately 22 percent of freshmen who entered a 4-year college or university in 2006 reported the intention to major in STEM;[6] in 2009, only about 16 percent of degrees were in a STEM field.[7] This phenomenon is most notable in engineering and, to a lesser extent, the physical sciences.[8]
- More than 50 percent of the doctorates awarded in the years 2006-2009 were in a STEM field, about a 9 percent increase from the beginning of the decade.[9]
- Among employed people in 2006 who had graduated in academic years 2003-2005 with a bachelor's degree in a STEM field, about 63 percent were in a STEM or STEM-related occupation; the comparable number was roughly 81 percent for those graduating with a STEM master's degree[10] and even higher for those at the doctoral level.

The U.S. economy is becoming more dependent on STEM workers. Indeed, as noted in Chapter 3, STEM occupations are projected to grow slightly faster than other occupations. Underrepresented groups such as women and non-Asian minorities are potential target groups for increases in the STEM workforce; STEM occupations pay above the average for these groups, and adding them would increase diversity.

Indeed, recent data indicate the following:

- Women accounted for approximately 57 percent of all bachelor's degrees earned over the years 2000-2009. The percentage of bachelor's degrees women earned in STEM fields during this period ranged betwen 10 and 11 percent; for men, however, the percentage of degrees earned in STEM fields ranged between 23 and 25 percent, more than twice the rate for women.[11]
- Although the share of African-Americans and Latinos in the overall pool of college students has been growing over the past 3 decades to about 26 percent of all undergraduates (including those seeking a 2-year degree), they still account for less than their 33 percent share of the college-age population would imply. Moreover, minorities (other than Asians) are even more underrepresented in STEM fields. While the overall percentage of 24-year-olds

[3]The other broad categories under consideration are physical sciences; mathematics/statistics; computer sciences; and engineering. Not included are social/behavioral sciences. See Appendix Table 2-12 in National Science Board (2012).

[4]Computer science, narrowly defined, is a relatively small field compared to other degree fields leading to employment in computer-related occupations such as electrical engineering. Recent data (see IPEDS) indicates that bachelor's degrees in computer science are once again rising. See Appendix Table 2-18 in National Science Board (2012).

[5]See Appendix Table 2-19 in National Science Board (2012).

[6]See Appendix Table 2-12 in National Science Board (2012).

[7]See Appendix Table 2-18 in National Science Board (2012).

[8]Note, however, that an examination by Xie and Killewald (2012) of three cohorts of high school seniors (1972, 1982, and 1992) found that "there is little evidence that science suffers from a 'leaky pipeline' during the college years that disproportionately steers students away from scientific fields." Moreover, according to Xie and Killewald, "teenagers' expectations of their future educational outcomes are full of noise" and "many students shift into and out of science, especially around the time of entering college." Further information is available in Xie and Killewald (2012).

[9]See Appendix Table 2-27 in National Science Board (2012).

[10]See Tables 35 and 36 in National Science Foundation (2010).

[11]In addition, from 2000 to 2009, the share of all bachelor's degrees awarded to women declined in computer sciences (by 10 percentage points), mathematics (by 5 percentage points), and engineering (by 2 percentage points). See Appendix Table 2-18 in National Science Board (2012).

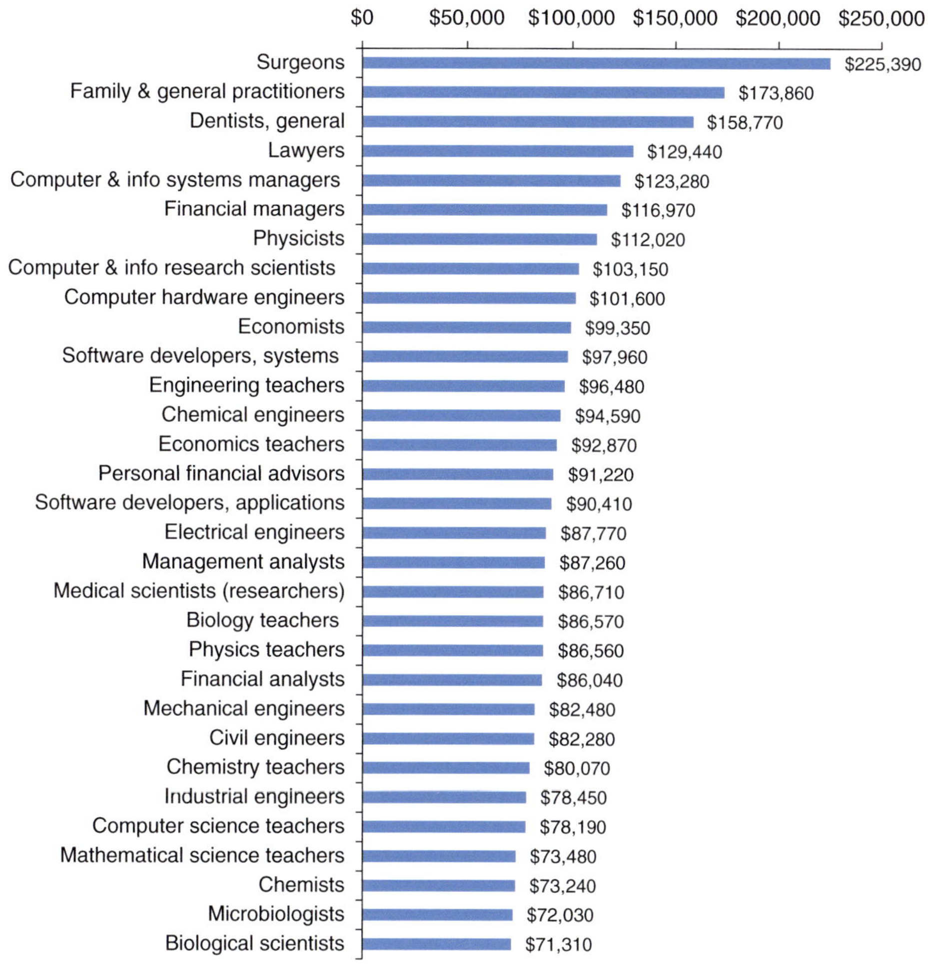

FIGURE 4-1 Annual wage estimates for select occupations, May 2010.
SOURCE: Bureau of Labor Statistics, U.S. Department of Labor, Occupational Employment Statistics, www.bls.gov/oes/.

in the United States holding a STEM degree is 6 percent, it is only 2.7 percent among African-Americans and 2.2 percent for Latinos (Mervis, 2010).

Not all indicators on the flow of talent into the STEM pipeline are promising:

- While there has been only a slight decline since 1977 in the percent of high school graduates who go on to complete or enroll in a STEM field in college, the percentage of "talented" students (defined as the top quintile on the ACT or SAT) doing so peaked for the 1992/1997 cohort and fell by almost 50 percent for the 2000/2005 cohort, suggesting that these "talented" students are being attracted to degrees and careers other than STEM (Lowell et al., 2009).[12]

[12]STEM includes the life and physical sciences, engineering, mathematics and information technology, and science and engineering technicians (and excludes the social sciences).

- While the retention of STEM graduates in STEM occupations 10 years after high school graduation rose, on average, from 34.8 percent for the 1977/1987 cohort to 43.7 percent for the 1993/2003 cohort, retention in STEM occupations for the most-talented group seemed to decline, although the decline was not statistically significant.[13]

These data raise questions about the perceived attractiveness of STEM occupations relative to others available to talented individuals. One issue is relative salaries. Evidence suggests that especially for males and U.S. citizens, relative salaries do have a bearing on who does science (Stephan, 2012, pp. 5, 153-156). As data from the Bureau of Labor Statistics (BLS) in Figure 4-1 (on the preceding page) show, with the exception of some IT-related occupations, jobs in management, finance, the medical professions (primarily, medical doctors and dentists), and law typically pay more on average than STEM occupations in either industry or academe. Moreover, postsecondary faculty positions in STEM fields often require many more years of education and training than for these other occupations, with the exception of some medical specialties.

Furthermore, the earnings profile over a career varies by occupation. For example, according to Stephan (2012), early-career PhD engineers (i.e., those who have had their doctorate for less than 10 years) earn about 1.6 times more than those with a bachelor's degree in any field who are aged 25-34. Early career PhD physical scientists earn about 1.4 times more, whereas early career PhD life scientists earn less than 1.3 times more (Stephan, 2012, p. 154). The picture is no better when one examines the relative earnings of late career scientists, that is, those 10 to 29 years into their career.

Taking into account the cost of obtaining a PhD in terms of earnings forgone during the years pursuing the degree and subsequent years of training as a postdoctoral fellow, Stephan (2012; p. 157) estimates that the present value of an MBA degree (which typically takes no more than 2 years to complete) is, on average, about $3.2 million dollars, while the present value of the PhD (which often takes 7 or more years to complete) is much lower at about $2 million dollars (Stephan, 2012). Furthermore, Stephan finds that those with MBAs from the best programs can expect to earn (over their lifetime) four to five times more than the average MBA,[14] while PhDs hired at top research universities can expect to earn (over their lifetime) only about three times more than the average PhD (Stephan, 2012).

The Role of Temporary Residents in Meeting STEM Needs

Using data on the composition of the STEM workforce based on the 2000 decennial census (National Survey of College Graduates 2003), Table 4-1 adapted from Levin and Barker (2010) shows the initial entry[15] visa of STEM[16]-educated (by highest degree earned) migrants in the United States as of 2003 by birth region and entry cohort. Several trends are evident:

- The entry visa types have changed over time, with temporary visas now outnumbering permanent visas (i.e., green cards). See Figure 4-2.
- The country of origin of these migrants has changed dramatically over the decades, with migrants from Asia, especially from China and India, growing much faster than migrants from Europe. See Figure 4-3.
- Among the temporary visas types, temporary work visas (primarily H1-B) have grown the fastest and are now nearly as plentiful as temporary study visas. See Figure 4-2. Nonetheless, commercial firms continue to cite the lack of H-1B visas as a significant problem in hiring needed talent.

[13]The change in retention from 44.8 percent to 43.2 percent.

[14]At least temporarily, the financial crisis has dampened the expected returns to careers in finance.

[15]For a period of at least 6 months.

[16]Excludes those educated in the social sciences.

TABLE 4-1 STEM-Educated Migrants in the United States in 2003 by Birth Region (Country), Initial Entry Visa Type, and Cohort

Region/Country	All Visas #	Green Card %	Temporary Work %	Temporary Study %	Temporary Depend. %	Temporary Other %
			Before 1970s			
All	315,452	51.3	1.4	28.6	13.0	5.7
Europe	97,041	66.7	1.9	11.6	16.6	3.2
Asia	118,692	31.2	1.3	52.7	10.9	3.8
China	19,819	27.9	1.0	58.5	7.4	5.3
India	21,186	25.6	1.0	67.4	5.7	0.4
			During 1970s			
All	456,977	48.4	6.4	29.1	9.2	6.9
Europe	52,162	54.4	10.3	15.8	8.5	11.0
Asia	308,797	49.2	6.2	31.2	8.6	4.8
China	9,792	37.1	0.0	56.6	4.6	1.7
India	67,218	58.2	3.0	28.2	9.6	1.0
			During 1980s			
All	645,561	42.5	9.3	31.9	7.2	9.1
Europe	93,985	37.8	18.6	20.4	8.7	14.4
Asia	408,471	44.9	7.1	34.8	6.7	6.4
China	58,680	23.5	1.8	62.6	10.9	1.2
India	83,128	50.5	4.4	35.3	4.7	5.2
			During 1990s			
All	896,143	26.7	25.1	28.6	10.9	8.7
Europe	186,038	38.4	19.3	21.7	6.7	13.8
Asia	525,352	21.7	26.6	32.7	14.1	4.8
China	110,746	7.6	14.6	57.1	17.6	3.0
India	205,917	19.6	36.1	26.6	14.6	3.1

Note: Numbers are subject to rounding errors.
SOURCE: Adapted from Levin and Barker (2010).

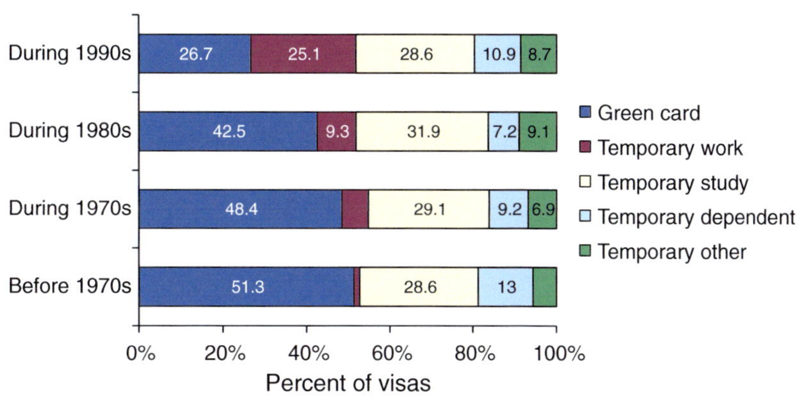

FIGURE 4-2 STEM-educated migrants in the United States in 2003 by initial entry visa type and cohort. NOTE: Numbers are subject to rounding errors.
SOURCE: Levin and Barker (2010).

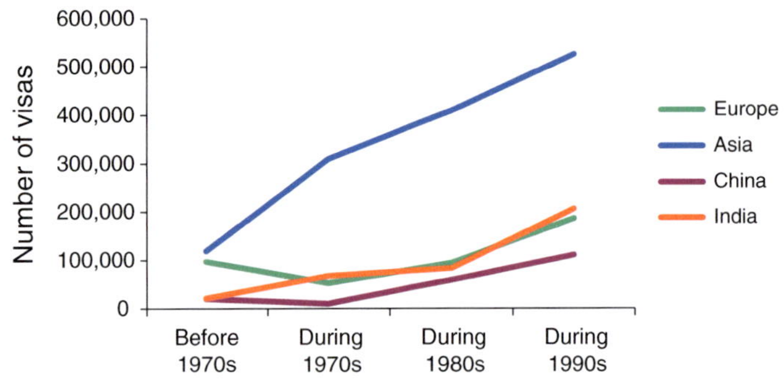

FIGURE 4-3 STEM-educated migrants in the United States in 2003 by birth region (country) and cohort. NOTE: Numbers of visas for Asia includes those for China and India.
SOURCE: Levin and Barker (2010).

Temporary Work Visas

H-1B visas are an important vehicle by which migrants enter the STEM workforce. Such visas likely account for the largest number of highly skilled workers who are entering the country with temporary work visas.[17] These visas are typically issued for 3 years and can be renewed for an additional 3-year period. The visa was started in 1990 with a cap of 65,000 per year; in 2001 the cap was tripled to 195,000 per year for 3 years but has now returned to its original level. Universities and non-profit research institutions are exempted from the numerical caps entirely (Wasem, 2012). Starting in 2005, an additional 20,000 visas were granted to students who had received master's degrees or doctorates from U.S. schools and were thus exempt from the cap of 65,000 (P.L. 108-447). In 2010, the United States issued more than 118,000 H-1B visas. This was down almost 25 percent from the nearly 155,000 issued in 2007, but this is likely only a temporary downturn due to the poor economy in the United States. The available data suggest that most H1-B visa recipients work in science and engineering (S&E) and S&E-related occupations. In 2009, 35 percent of new H-1B visa recipients were employed in the category of computer-related occupations.[18]

Considering educational attainment, in FY 2009, 58 percent of new H-1B visa recipients had an advanced degree, including 40 percent with master's degrees, 6 percent with professional degrees, and 13 percent with doctorates. The distribution by degree-level varies by occupation, with 83 percent of mathematical and physical scientists holding advanced degrees (44 percent with doctorates). Among life scientists, 87 percent hold advanced degrees (61 percent with doctorates).[19] It is likely that a substantial number of those with PhD degrees are in relatively low-paid postdoctoral positions at U.S. universities.

Overall, these data demonstrate both the use of the H-1B visa as a way for foreign graduates of U.S. schools to undertake postdoctoral training or otherwise pursue careers in the United States, at least temporarily, as well as the importance of the H-1B visa in bringing foreign-educated individuals to the United States, especially in STEM occupations.

The use of H1-B visas to meet STEM workforce needs in the United States is, however, a continuing source of controversy. Industry argues that these workers are meeting shortages of workers with particular skills; others argue that the inflow of these workers may be discouraging U.S. citizens from pursuing education and jobs in these skill areas (Levin et al., 2004). Likely, the answer lies somewhere in between these two positions. It is doubtful that

[17]Other categories of temporary work visas include the J-1 Exchange Visa, which is often given to lower-skilled workers and summer visitors and the L-1 Visa issued for intracompany transfers. The latter category has been growing very rapidly and from 2006 to 2010 averaged about 76,000 annually. See Figure 3-36 in National Science Board (2012). Here the NSF definition of S&E is utilized.

[18]See Appendix Table 3-19 in National Science Board (2012).

[19]See Chapter 3 in National Science Board (2012).

DOD's special concerns about the percentage of STEM graduate students who are "clearable" would be influential in resolving such a debate. Changes in H1-B visa policies that largely affect STEM occupations will have repercussions for the pool of highly skilled applicants available to DOD and its industrial base. In the short run, further constraints on H1-B visa entrants may make it more difficult for DOD to recruit citizens if these constraints increase competition for them in the private sector. In the longer run, however, if market forces cause wages to increase following a tightening of H-1B visa policy, more citizens may eventually pursue careers in STEM occupations.

Sandia National Laboratories has a hiring pathway by which a foreign national can become a member of its technical staff. The first stage for the prospective staff member is to become established as a staff member (e.g., in a postdoctoral position or as a limited-term employee). Next they are converted to Foreign National Interim Technical Staff, which includes a requirement that they concurrently pursue a path to U.S. citizenship. Due to the classified nature of Sandia's work, the prospective staff member must obtain the necessary security clearances and successfully pass a comprehensive counterintelligence investigation. Upon completion of the latter, or receipt of citizenship, the individual becomes a member of the technical staff.

Temporary Study Visas

Non-citizens (primarily with temporary study visas) also play an important role in the production of S&E degrees in the United States, primarily at the master's and doctoral level.[20] In 2009, foreign students earned 38 percent of U.S. S&E master's degrees. In computer sciences and engineering, however, they earned 46 percent and 43 percent, respectively, of all such degrees.[21] And within engineering, they earned more than half of the master's degrees in electrical and chemical engineering. At the doctoral level, temporary residents earned 35 percent of all S&E degrees awarded in 2009. But they accounted for 57 percent of doctoral degrees awarded in engineering, 44 percent in physical sciences, and 54 percent in computer sciences, although only 29 percent in the biological sciences and 8 percent in medical/other life sciences.[22]

A large number of these temporary residents, especially at the doctoral level, stay in the United States for at least 5 years after graduation, although the numbers vary by source country. Analysis of data from the Social Security Administration (Finn, 2012) shows an average 5-year stay rate of 62 percent in 2009 for temporary residents receiving a science or engineering doctorate in 2004, with China and India having the highest percentages at 89 percent and 79 percent, respectively. For the 1995, 1997, and 1999 cohorts of foreign national science and engineering doctorate recipients, the stay rates tend to fall slightly from 6 to 10 years after graduation, although stay rates for these cohorts are considerably higher than the 10-year stay rates of earlier cohorts (1991, 1993). While there are tremendous differences in stay rates by source country, these have remained stable over time. Moreover, despite media reports of a "brain drain" of foreign scientists and engineers out of the United States,[23] Finn (2012) states that "stay rates are more likely to increase in coming years than to decline" because (1) the share of foreign science and engineering doctoral degrees recipients coming from countries with the highest stay rates has been increasing and (2) those intending to stay in the United States after graduation as reported in the Survey of Earned Doctorates have increased since 2004 (p. 14). Finally, in an earlier report, Finn (2010) posits that the performance of the U.S. economy may affect stay rates, although stay rates declined only modestly during the recession of the early 2000s.

Security Clearances

DOD and its associated contractors have special and legitimate needs to hire STEM personnel who can obtain security clearances. Under current practices this generally requires U.S. citizenship, and special problems therefore can arise in hiring in STEM fields in which large proportions of students at U.S. universities are foreign nationals.

---

[20]At the bachelor's level, non-citizens account for less than 5 percent of the degrees awarded, although in 2009 they accounted for about 9 percent of the degrees in electrical and industrial engineering. See Appendix Table 2-19 in National Science Board (2012).

[21]See Appendix Table 2-26 in National Science Board (2012).

[22]See Appendix Table 2-28 in National Science Board (2012).

[23]See, for example: Herbst (2009); Lee (2011); Wadhwa (2011).

Security clearances are typically classified at one of three levels: confidential, secret, or top secret. Gaining access to sensitive compartmented information or special access programs can also necessitate a top secret clearance. DOD-issued clearances constitute the vast majority of initial clearances (Government Accountability Office, 2010). In the context of the pool of STEM workers available to DOD, the need to obtain a security clearance is a two-fold source of constraint on supply. First, the time required to obtain a security clearance for citizens represents an impediment to success in DOD's hiring process. Second, this requirement reduces the pool of the potential STEM workforce for DOD in fields in which non-citizens represent substantial fractions. While this is not much of a problem at the BS level in engineering, where international students represent only a small percentage,[24] it is a significant problem for positions requiring graduate engineering degrees, where the percentage of temporary residents, as noted earlier, is much higher.

A recent study of personnel security clearances found that progress has been made in reducing the time to adjudicate applications (Government Accountability Office, 2010). Specifically, the report noted that DOD was able to meet the goal of adjudicating 90 percent of its applications within 60 days. The process also allows for the DOD to give interim clearances at the secret level once an investigation on an individual has been opened and no initial problems have been identified (Department of Defense, 1999; Secretary of the Navy, 2006). (In contrast, temporary access at a top secret level can be granted only if the applicant already has a secret or a confidential level clearance.) Otherwise, a secret level interim clearance can be given to those requesting top secret level access and only if a local review of the personnel security questionnaire (PSQ) is found to reveal no eligibility issues. Those DOD commands that do not impose restrictions for facility access according to security clearances could potentially hire STEM researchers, with the caveat that their continued employment requires a security clearance at the appropriate level. Despite recent improvements, it is possible that some top STEM talent, who can be hired on the spot by private-sector recruiters, may be deterred from pursuing DOD careers by delays in their appointments because of clearance issues.

The system of personnel security clearances is far from being the only set of controls placed on those performing defense work or on the goods and services they produce, including their export or so-called deemed export. A technology on the U.S. Munitions List, administered by the U.S. Department of State, is subject to export controls (Congressional Research Service, 2009). Another set of controls applies to so-called dual-use technologies on the Commerce Control List (CCL), administered by the U.S. Department of Commerce. Activities conducted within the United States, such as sharing knowledge of a technology with a foreign national residing domestically, may constitute a deemed export, requiring a license or exemption under the International Traffic in Arms Regulations (National Research Council, 2009a, p. 34). The system of controls and their implementation is formidable, and interested readers are urged to consult the substantial secondary literature on this topic (Center for Security and International Studies, 2005).

Unfortunately, U.S. government policies regarding the funding of higher education (and particularly graduate education) in STEM fields lack coordination with policies regarding temporary visas for education and work as well as for visas for permanent residence. For example, large amounts of financial support from federal agencies—via research grants—are used by U.S. universities to finance the graduate education of international students in STEM fields. Immigration policies allow universities essentially unlimited access to such students, as well as to international "postdocs," i.e., those who have earned PhDs in countries other than the United States and then come to work as postdocs in U.S. university labs with financial support from federal research grants. Surprisingly there are almost no credible data on this apparently large and growing population of international postdocs, though it appears that the largest country of origin is China. Meanwhile, as noted earlier, more than 100,000 temporary skilled workers each year are admitted with H1-B visas, mostly in STEM and related IT fields, for temporary but multi-year work. Yet the number of permanent visas available based on these same skills is much smaller, resulting in large backlogs of temporary visa holders seeking permanent visas.[25]

[24] In 2009, only 5.8 percent of the undergraduate degrees in engineering were earned by temporary non-resident students. See, Appendix Table 2-19 in National Science Board (2012).

[25] It now can take as long as 6 to 10 years to obtain the coveted "green" card that grants permanent residency to skilled-immigrant workers from China and India, for example (Wadhwa et al., 2007). Moreover, Hira (2010) argues that "most of the top users of both the H-1B and L-1 visa programs sponsor very few, if any, of their workers for permanent residence."

DEMAND-SIDE ISSUES

The DOD must compete with the civilian sector for job opportunities available to STEM-trained individuals. Data provided by the Bureau of Labor statistics (BLS) can help gauge the strength of competing demands from the civilian sector. BLS provides data on employment and wages, including employment projections by occupation (see Chapter 3 of this report). It should be noted, however, that in addition to the array of occupations normally included in STEM employment numbers—engineers, math and computer scientists, and life and physical scientists—BLS included STEM technicians, architects, postsecondary teachers in STEM fields, STEM managers, and those in STEM-related sales jobs in a recent study of the STEM workforce (Cover et al., 2011, pp. 3-15).

- As reported in Chapter 3, of the 7.3 million employed in STEM in the civilian sector in 2010 (accounting for 5.1 percent of the overall workforce), the greatest number is employed in computer-related occupations.[26]
- For the period 2010 to 2020, STEM employment is projected to grow by 16.9 percent, which is slightly higher than the projected growth rate of 14.3 percent for the workforce as a whole.

Although these are projections premised on assumptions regarding future GDP growth which are themselves subject to considerable uncertainty, they do at least suggest relatively strong growth in the civilian sector in those occupational categories likely to be most sought after in the coming years by DOD and the industrial base. Moreover, Sauermann and Roach (2012) found in a survey of PhD students that "a faculty research career is the career path most often considered 'extremely attractive' and ranks among the most desirable careers for over 50% of life scientists and physicists," suggesting that DOD will continue to face competition from academic institutions for PhD-level scientists. There are several other important issues that DOD and its industrial base must confront in order to meet its STEM workforce needs.

Pay

The DOD's ability to pay uniformed, civilian, and, indirectly, contractor STEM workers competitive salaries will be a further issue to consider in developing DOD's STEM talent. Data from the Congressional Budget Office and the Project on Government Oversight suggests that STEM workers above the bachelor's level are paid less in the civilian federal workforce than in the private sector (POGO, 2011). For example, CBO finds that individuals in the federal workforce with a professional or doctoral degree earn (in wages and benefits) about 18 percent less than their counterparts in the private sector (Congressional Budget Office, 2012). Other federal agencies such as NIH, NSF, and EPA have Title 42 authority "to appoint highly qualified consultants, scientists and engineers at a pay scale [up to $250,000 per year] outside civil service laws described under Title 5" (National Research Council, 2010). Salaries may impact the STEM pipeline by providing a signal to prospective STEM majors.[27] For example, the number of persons graduating with a degree in computer science increased, with a lag of a few years, as wages for computer programmers increased in the early 1990s. Furthermore, the number of persons graduating with a degree in computer science declined as wages stagnated in the early 2000s (Figure 4-4). Similarly, Ryoo and Rosen (2004), in their examination of the engineering labor market, found that engineering enrollment decisions appeared to be sensitive to engineering career prospects (as measured by the present discounted value of earnings in engineering relative to alternative professions). Lastly, a study of science and engineering PhD students by Roach and Sauermann (2010) found that students "concerned with salary, access to resources, and the desire to conduct downstream research and development" are more likely to prefer a career in industry over a career in academia. These results suggest that pay is an important aspect of the value proposition the DOD can offer to a prospective employee.

That said, extended discussion with a senior representative from DOD's Office of Personnel and Readiness-suggested that there is no shortage of qualified applicants for the positions advertised on the DOD website and in

[26]Postsecondary STEM teachers are excluded from the estimate of STEM employment since the information available from the Bureau of Labor Statistics' Employment Projections Program does not distinguish teaching field.

[27]Freeman (1976) established that "the supply of new entrants to engineering is highly responsive to economic conditions."

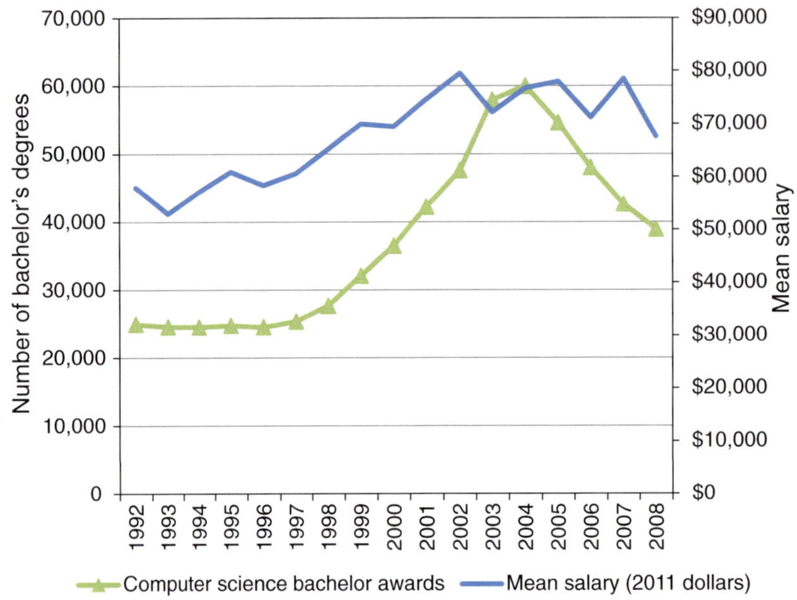

FIGURE 4-4 Computer science bachelor's degree awards and computer programmer real mean salaries, 1992-2008.
SOURCE: Kuehn and Salzman (2013).

trade papers.[28] This is particularly true in the current job climate where available positions are scarce. The committee was advised that applicants come because they find attractive the opportunities for greater responsibility in the government labs during the first 5 years of work than there would be in private labs. DOD entry level compensation was declared to be sufficiently attractive, particularly when one includes bonuses. Another DOD source noted that at the top end of the salary scale, somewhere around $160,000, the private sector enjoys a distinct advantage since the government is not competitive. In these circumstances, one has to be concerned with the appropriateness of the skills of the persons being hired and the potential deleterious impact on activities.

Quality of Work

To attract top talent, the work DOD and its industrial base offers must offer sufficient challenge and importance to excite the most creative and highly skilled workers, and to motivate them to achieve peak performance. "Pay for performance" personnel policies can be implemented, but if the work is not sufficiently exciting, pay alone will not be enough.

In the past, major DOD procurement programs have been a sufficient source of widely visible program development challenges, attracting and motivating the talented workforce DOD desired. However, major DOD procurement programs are decreasing. For instance, there are currently only two major aircraft programs in development, following a steep decline in numbers of new starts since the Second World War (Figure 4-5). There are, however, exciting smaller-scale programs in DOD in a number of areas that may be less visible to the general public. For instance, the Defense Advanced Research Projects Agency (DARPA) continues to support advanced concept technology demonstrations across a spectrum of disciplines.[29] The Special Operations Command sponsors cutting-edge field experimentation in the academic environment of the Naval Postgraduate School. These types of programs tend to precede the competitive phase of the acquisition process, which leaves them relatively free of

[28] Pasquale "Pat" Tamburrino, deputy assistant secretary for civilian personnel policy, personal communication.

[29] Note, however, that some have argued that DARPA has become too risk averse. See, for example, Ignatius (2007).

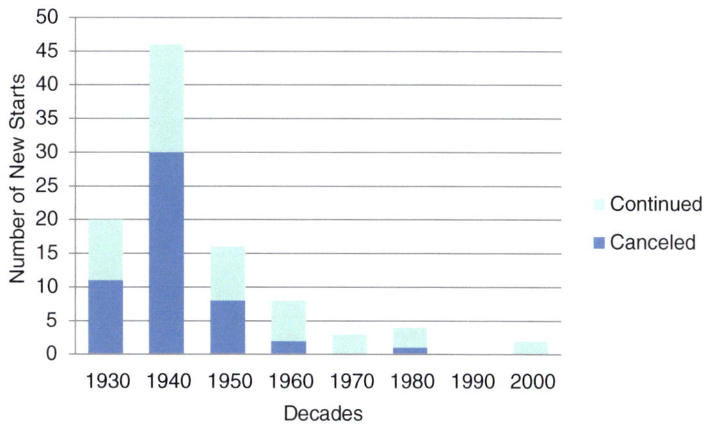

FIGURE 4-5 Number of new fighter and bomber starts per decade. SOURCE: Carlson and Chambal (2008).

bureaucracy. Innovation and creativity are encouraged. High-risk projects are allowed to "fail," and researchers, practitioners, and students are encouraged to push the envelope.

Another such example is the Rapid Reaction Technology Office (RRTO), later folded into the Rapid Fielding Office within ASD(R&E), which was charged with developing counterterrorism technologies and employed rapid prototyping. DOD established the RRTO in 2006 in response to the constantly evolving threat of asymmetric warfare, including, for example, the use of improvised explosive devices (IEDs) in the Iraq and Afghanistan theaters of operation. Established under the director, defense research and engineering, it focused on developing technologies that can mature in 6 to 18 months for the purpose of countering insurgency and irregular warfare. The RRTO provides a diverse set of quick-response capabilities for counter-terrorism while attempting to stimulate interagency coordination and cooperation. The office operates without a formal charter or governing document, and the director has much flexibility for carrying out the mission. Approximately 50 percent of the office's projects have resulted in fielded technologies, altered concepts of operation (CONOPS), or other concrete changes in larger systems. Such projects included the Persistent Threat Detection System for persistent ground surveillance through a tethered aerostat with an embedded camera; a Biometric Automated Toolset for screening personnel in mobile applications; and the SKOPE intelligence cell, a joint analytic cell with the National Geospatial Intelligence Agency, U.S. Special Operations Command, and U.S. Strategic Command. Strategic investment by DOD in programs of this nature appears to be an important cornerstone of creating an increasingly attractive workforce environment (National Research Council, 2009b).

In a similar vein, the Lockheed Martin Skunk Works® has, over its nearly 70-year history, created breakthrough technologies and landmark aircraft that continually redefine flight. Guided by the mantra "quick, quiet, and quality," the Skunk Works® requires a flexible workforce capable of quickly forming and disbanding interdisciplinary project teams. To meet this need, the Skunk Works® uses a matrix organization that minimizes paperwork and delays in moving people between teams. Core engineering groups maintain skill sets and tools to support their disciplines. Program managers draw their teams from these talent pools. Likewise, NASA developed its Engineering and Safety Center (NESC) in 2003 to provide an independent test, analysis, and assessment capability to NASA programs and projects. The NESC operates independently of mission directorates and reports to the Office of the Chief Engineer. The NESC operates through technical discipline teams (TDTs), each led by an agency-recognized NASA Tech Fellow, who is an outstanding senior-level engineer or scientist with distinguished and sustained records of technical achievement. The fellows provide leadership and act as role models for NASA discipline engineering communities beyond the TDTs which are drawn not only from NASA but also from other federal agencies, industry, and universities, making them diverse teams that can provide robust, creative solutions to complex problems. Another government agency supporting high-risk ventures is the Advanced Research Projects Agency-Energy (ARPA-E), which funds specific high-risk, potentially high-payoff, energy research and development projects. ARPA-E has been set up to be a lean and agile organization with special hiring authority to bring on program directors and other program leadership with the ability to offer limited-term rotational assignments. Thus, individuals from all sectors are able to assume temporary positions lasting roughly 3 years. The agency empowers them to make technical and programmatic decisions for the projects they oversee (PCAST, 2010; Yehle, 2011).

Until 1998, the DOD budget included a category "6.3B" for systems advanced development that supported rapid prototyping programs (National Research Council, 2001).

Quality of Workplace

For any organization seeking to maximize the productivity of its professionals in science and engineering, high-quality, up-to-date facilities and equipment are essential. In addition, the availability of such facilities and equipment enhances the recruitment of talented scientists and engineers.

The committee is aware of a series of reports that describe limitations experienced by DOD research labs in this regard.[30] However, the most recent such report is already several years old, and it appears that there was little action in response to the recommendations during subsequent periods of budgetary stringency. The committee agrees with the perspectives expressed in these reports. For example, a recommendation from a 2001 report noted that DOD "should continue to pursue world-class status for the Service laboratories" and emphasized that this should be done "not only to obtain the highest-quality results from its research, but also to attract superior scientific and engineering personnel who want to work where the best research is done" (National Research Council, 2001). DOD's need for outstanding science and engineering in support of its increasingly technical missions does require that serious attention be paid to ensure that facilities and equipment available to DOD scientists and engineers are of the highest quality.

Work Environment

In order to attract the highest-quality workers, the DOD should consider personnel policies as they relate to the ability of DOD to attract, retain, and develop the STEM workforce it needs. In a July 2010 study entitled *Defense Acquisition Workforce Modernization* from the Center for Public Policy and Private Enterprise at the University of Maryland, the authors posit that "to effectively develop the required human capital for the modern acquisition environment, we believe that DOD should enhance its recruitment processes; improve the hiring process; strive for quality not quantity; provide compelling wages; incentivize employees for improved performance; and, incentivize employees for additional training and education" (Gansler et al., 2010). These imperatives can be generalized to the STEM workforce as a whole. For example, sabbatical and expanded internship programs, as well as online and anytime/anyplace programs, address not only recruiting and retention issues but also the increasingly interdisciplinary competencies required by the workforce.

Global Competition

DOD will be have to compete for scientific talent in the changing environment of globalization. Witness the growth in higher education and the development of technological infrastructures in S&E in China and India, two large suppliers of U.S. STEM graduates. In these countries, there is going to be a demand for these graduates that did not exist previously. Even Russia, whose scientific enterprise has suffered from the migration of scientific talent since the breakup of the Soviet Union, has witnessed the appointment of a U.S. engineer as president of the new graduate research university—the Skolkovo Institute of Science and Technology—a collaborative effort with MIT.[31]

[30]For more information see, National Research Council (1990, 2001, 2005). See also JASON (2009), p. 26.

[31]According to press releases from MIT, "This institution aims to break new ground in bringing together Russian, US and global research and technology—and in integrating teaching, research, innovation and entrepreneurship" (MIT News, 2011).

Furthermore, some countries are trying to attract back home that segment of their native-born scientific talent that has been educated in the United States.[32] Small countries such as Singapore are luring scientific talent from the best universities and labs in the world in order to build a cutting-edge science enterprise that is intended to transform Singapore "into a knowledge-based, innovation-driven economy."[33] China and Taiwan, among others, are also actively seeking foreign talent and are now rapidly developing their own technological infrastructures, making STEM careers in such graduates' home countries increasingly attractive (Wadhwa et al., 2009). In this more competitive environment, one would expect that the quality of the research environment as well as the remuneration paid to scientists and engineers will become increasingly important, particularly for foreign students who have come to the United States seeking economic success.

Recruiting—Increasing Public Awareness

To counter these competitive pressures, DOD should offer highly competitive career opportunities for outstanding scientists and engineers. DOD advertisements for STEM applicants may be neither visible enough nor attractive enough in conveying the exciting research underway under DOD auspices. Because these jobs are about national security, they should be seen and advertised as critical to the defense of the nation, thus appealing to patriotic instincts. Awareness efforts could be informed by advertising campaigns such as those that have been developed by the Marines. These campaigns appear to be effective in creating a sense of purpose (the defense of the nation, referencing "the Marines"), exclusivity (referencing "the Few"), and a profound sense of superiority (referencing "the Proud"). DOD scientists and engineers are not uniformed combatants, but their work is an essential part of our national security mission and the U.S. role in promoting global peace. A concerted effort on the part of DOD to bring awareness to the vast contributions of its highly diversified STEM workforce could go a long way toward moving DOD into the vanguard of crafting a "heroic" image for the agency scientists and engineers whose work is vital to U.S. national security. This was in fact a major factor in attracting talent during the Cold War era.

The committee knows of no DOD recruiting effort for civilian scientists and engineers that is comparable to those for the uniformed services, and yet there could be substantial commonalities with the military system. DOD scientists and engineers play central roles in creating the tools with which the military service members operate on a daily basis. The development of a sophisticated civilian recruiting effort that identifies DOD scientists and engineers as working closely with military personnel in ways that are critical to national security could be highly effective. Moreover, many technically oriented students are attracted to intriguing and unique applications of science and technology, some of which are being led by DOD. These include globally controlled, unmanned aerial systems (UASs); "smart" weapons; sophisticated night vision; and the integration of complex communication and data that can be deployed in real time in battlefield conditions. Outreach and recruiting efforts could be amplified by offering highly qualified young science and engineering students internship opportunities in R&D in appropriate DOD labs, thereby exposing them to the exciting science and engineering challenges faced by the DOD.

There is no question that STEM disciplines will continue to grow in importance as defense capability becomes more technology-driven.To respond to this, one possibility would be to create a specialized recruiting function within the DOD that would be responsible for STEM recruitment and hiring. A second task of this office would be the identification of a list of higher education institutions that produce the students who best fit the demands of the workforce in the disciplines of the greatest interest to DOD (e.g., civil engineering, electrical engineering, petroleum engineering, etc.). A place to start would be with institutions that currently offer courses at DOD facilities and others with which cooperative structures exist, for instance, the Community College of the Air Force, eArmyU, and others.

---

[32]See for example, Lim (2011); Sharma (2011).

[33]The Agency for Science, Technology and Research (A*STAR) is the lead agency for fostering world-class scientific research and talent for Singapore. A*STAR oversees 14 biomedical sciences and physical sciences and engineering research institutes, and six consortia and centers, located in Biopolis and Fusionopolis as well as their immediate vicinity. It also supports educational programs in S&E at all levels of instruction. For more information, see http://www.a-star.edu.sg/.

BROADER ISSUES THAT MAY IMPACT DOD'S STEM WORKFORCE

In addition to the issues noted above, there are several exogenous factors that may have an impact on DOD's ability to hire and manage an effective STEM workforce, including the following:

- *The failure of the congressional "Super Committee" to reach agreement on budget cuts, which will likely result in another massive reduction in the DOD budget in the coming years.* While the White House and Congress agreed last summer on $487 billion in cuts to defense spending over the next 10 years, even deeper cuts are threatened if Congress fails to pass a new plan for deficit reduction. In that case, the Pentagon budget will be cut by a total of roughly $1 trillion over a decade, beginning in January 2013.[34]
- *The history of large swings in DOD funding.*[35] Defense spending increased sharply to over 9 percent of GDP in the mid-1960s as U.S. involvement in Vietnam expanded. After large-scale withdrawal from Vietnam began in1969, defense spending as a share of GDP fell to less than 5 percent of GDP by the end of the next decade. The Soviet invasion of Afghanistan prompted an increase in defense spending to about 6 percent of GDP during the early 1980s. After the Berlin Wall was opened in November 1989 and communist governments in central and Eastern Europe collapsed, defense spending as a share of GDP dropped to the historically low level of about 3 percent. Defense spending increased again to nearly 5 percent of GDP after the attacks on September 11, 2001, and the wars in Afghanistan and Iraq began. In the committee's experience, DOD has dealt with tightened budgets by reducing, often disproportionately, funding for workforce training and development. In addition, reductions in the STEM workforce seem to have been carried out in a manner having more to do with numbers and less with justification premised on impact to military capabilities or quality of the workforce.
- *The need for DOD to manage a potentially large increase in retirements when the recession ends and housing and securities markets rebound.* As reported in Chapter 3, about one-third of the DOD civilian STEM workforce is eligible to retire (see Figures 3-19 and 3-20). Moreover, this eligibility rate is more than double the estimated retirement eligibility rate of the defense industrial base workforce. While the actual rate of retirement is low for both workforces, DOD is likely at greater risk from future retirements. The need to recruit, develop, and retain highly skilled employees across both traditional and emerging STEM disciplines such as translational computing, autonomous systems, systems biology, innovative materials, and efficient manufacturing should be a DOD priority.

FINDINGS AND RECOMMENDATIONS

Finding 4.1. Stable funding for the recruitment and development of STEM human resources is essential to their effective management.

Recommendation 4.1. The DOD should fund STEM recruitment and development in a manner that facilitates stability, such as multi-year programming, "one color" of money for STEM related costs,[36] or funding based on a percent of total obligational authority. This would facilitate stability for long-term STEM investments and greater consistency across and within the services. In addition, DOD should require all services to justify, as part of the approval process, STEM-related manpower reductions in terms of impact on technology-based capabilities and, where appropriate, whether there has been sufficient return on investment from those who have recently completed postsecondary education paid for by the government.

---

[34]See for example, Barnes and Entous (2012).

[35]See for example, Austin and Levit (2010).

[36]Congress provides funds to the DOD in different appropriation accounts ("colors of money," a term of art used in day-to-day discussions within the DOD). DOD military personnel are paid from one account—the "MILPERS" account. DOD civilians, especially STEM-related civilians, can be paid from more than one account, such as the Operation & Maintenance (O&M) account or the Research, Development, Test, & Evaluation (RDT&E) account. These accounts are managed with different sets of rules including programming procedures, approval levels of reprogramming, and duration of funds. Therefore, civilians paid with different "colors of money" are funded differently in terms of both procedures and funding levels. This can cause significant disruptions and disparities across the services when it comes to employment programming, hiring, training, RIFs, awards, and so on.

Finding 4.2. The U.S. STEM workforce is heavily dependent on non-citizens. DOD will need to reassess its requirement for security clearances for many STEM positions along with the processes by which many of its systems are developed and procured.

Recommendation 4.2. The DOD should find creative ways to hire STEM-qualified non-U.S. citizen personnel to support and advance designated S&T activities. Consideration should be given to those aspects of programs that are not classified and those that could accommodate lower-level clearances. The process should be codified and repeatable to ensure a sufficient number of candidates under appropriate circumstances. It is understood that this could require both policy and legislative changes, including but not limited to adapting the H1-B program, and the issuance of exemptions under ITAR and other applicable laws and regulations.

Finding 4.3. The United States, including DOD and its industrial contractors, is competing in an ever-growing world market for top scientific and engineering talent. For the DOD to recruit top STEM talent in competition with commercial firms, universities, and others, it must commit to improving the STEM workforce environment. The DOD must become, and be perceived as, an attractive career destination for the most capable scientists, engineers, and technicians, who are in great demand in the global talent marketplace.

Recommendation 4.3. The DOD should strengthen its ability to recruit, educate, and retain top STEM talent by offering competitive salaries and a constructive work environment, providing challenging and interesting problems in the workplace, enabling existing talent to keep up with the newly emerging scientific trends, and providing opportunities for the retraining of its STEM workforce to meet changing scientific and technological needs.

Finding 4.4. Because of the increasing acquisition costs of major systems and continuing pressures on DOD budgets, the number and variety of major weapons being developed and fielded have shrunk significantly in recent decades. This dynamic has a dampening effect on recruiting for the DOD STEM workforce.

Recommendation 4.4. The DOD should support, wherever possible, experimental and rapid-prototyping programs that push the cutting edge of science and engineering, in order to both maximize new technology applications and to attract the best and brightest STEM workers.

Finding 4.5. The DOD has centers of excellence across its own institutions, but the quality and the modernity of both facilities and equipment vary widely, marginalizing DOD's ability to compete broadly for top STEM talent.

Recommendations 4.5. The DOD should establish high standards of quality for both facilities and equipment and fund them appropriately.

Finding 4.6. DOD's personnel policies with regard to recruiting, hiring, paying, retaining, and incentivizing additional training and education are not currently optimized for maintaining the best STEM workforce.

Recommendation 4.6. The DOD should consider changes in personnel policy that would enable it to move more nimbly to make competitive hiring offers in DOD-critical scientific and engineering fields. Some of these changes can be made internally within DOD. Where this is not currently possible, DOD should seek legislative and/or regulatory relief. The following changes warrant consideration by DOD:

- More active outreach and recruitment efforts, aimed at civilian hires, of needed scientists and engineers that emphasize the many exciting technologies that are being developed by DOD and their potential contribution to the nation;
- New measures to expedite recruitment offers for occupations in which DOD determines that it must compete with more nimble corporate recruiters;

- Additional authority to expedite security clearances needed for such positions, including permission for temporary hiring into non-sensitive roles pending confirmation of security clearance;
- Actions to protect or "ring-fence" science and engineering positions determined by DOD to be critical capabilities, thereby protecting the loss of such capabilities due to future RIFs and hiring freezes; and
- Further provisions to incentivize DOD scientists and engineers to seek additional continuing education and training in rapidly developing areas of science and technology.

REFERENCES

Austin, D.A., and M.R. Levit. 2010. Trends in Discretionary Spending. Washington, D.C.: Congressional Research Service.

Barnes, J.E., and A. Entous. 2012. Pentagon to lay out next year's budget cuts. Wall Street Journal, January 25.

Carlson, B., and S. Chambal. 2008. Senior leader perspective. Developmental planning: The key to future war-fighter capabilities. Air and Space Power Journal 22(1):3.

Center for Security and International Studies. 2005. Security Controls on the Access of Foreign Scientists and Engineers to the United States. Washington, D.C.: Center for Security and International Studies.

Congressional Budget Office. 2012. Comparing the Compensation of Federal and Private-Sector Employees. Washington, D.C.: Congress of the United States.

Congressional Research Service. 2009. The U.S. Export Control System and the President's Reform Initiative. Washington, D.C.: Government Printing Office.

Cover, B., J.I. Jones, and A. Watson. 2011. Science, technology, engineering, and mathematics (STEM) occupations: A visual essay. Monthly Labor Review 134(May):3-15.

Department of Defense. 1999. Personnel Security Program Regulation. DOD 5200.2. DOD, April 9. Available at www.dtic.mil/whs/directive/corres/pdf/52002p.pdf.

Finn, M.G. 2010. Stay Rates of Foreign Doctorate Recipients from U.S. Universities, 2007. Oak Ridge, Tenn: Oak Ridge Institute for Science and Education.

Finn, M.G. 2012. Stay Rates of Foreign Doctorate Recipients from U.S. Universities, 2009. Oak Ridge, Tenn: Oak Ridge Institute for Science and Education.

Freeman, R.B. 1976. A cobweb model of the supply and starting salary of new engineers. Industrial and Labor Relations Review 29(2):236-248.

Gansler, J.S., W. Lucyshyn, and M. Arendt. 2010. Defense Acquisition Workforce Modernization. University of Maryland: Center for Public Policy and Private Enterprise.

Government Accountability Office. 2010. Personnel Security Clearances: Progress Has Been Made to Improve Timeliness but Continued Oversight Is Needed to Sustain Momentum. Washington, D.C.: Government Accountability Office.

Herbst, M. 2009. Why the U.S. Is Losing Foreign Graduates. Available at http://www.businessweek.com/technology/content/mar2009/tc20090318_162454.htm (accessed October 3, 2012).

Hira, R. 2010. Bridge to Immigration or Cheap Temporary Labor? The H-1B & L-1 Visa Programs Are a Source of Both. EPI Briefing Paper. Document 257. Economic Policy Institute, February 17. Available at www.epi.org/publication/bp257/.

Ignatius, D. 2007. The ideas engine needs a tuneup. Washington Post, June 3.

JASON. 2009. S&T for the National Security. McLean, Va.: MITRE Corporation.

Kuehn, Daniel, and Harold Salzman. 2013. The labor market for new engineers. U.S. Engineers in the Global Economy. Richard Freeman and Harold Salzman (eds.). National Bureau of Economic Research, forthcoming.

Lee, B. 2011. Reverse Brain Drain in the U.S. Available at http://www.pbs.org/wnet/need-to-know/the-daily-need/reverse-brain-drain-in-the-u-s/11027/ (accessed October 3, 2011).

Levin, S.G., and A. Barker. 2010. Studying Foreign Talent in the Science and Engineering Workforce Final Report to the Alfred P. Sloan Foundation. Grant number 2008-5-29 SEW.

Levin, S.G., G.C. Black, A.E. Winkler, and P.E. Stephan. 2004. Differential employment P patterns for citizens and non-citizens in science and engineering in the United States: Minting and competitive effects. Growth and Change 35(4):19.

Lim, L. 2011. China Aims to Renew Status as Scientific Superpower. Available at http://www.npr.org/2011/08/01/138837512/china-aims-to-renew-status-as-scientific-superpower (accessed October 3, 2012).

Lowell, B.L., H. Salzman, H. Bernstein, and E. Henderson. 2009. Steady as She Goes? Three Generations of Students through the Science and Engineering Pipeline. Paper presented at Annual Meetings of the Association for Public Policy Analysis and Management, Washington, D.C., October 9.

Mervis, J. 2010. New Answers for Increasing Minorities in Science. Available at http://news.sciencemag.org/science insider/2010/09/new-answers-for-increasing-minorities.html (accessed October 3, 2012).

MIT News. 2011. Skolkovo Foundation and MIT to Collaborate on Developing the Skolkovo Institute of Science and Technology. Available at http://web.mit.edu/newsoffice/2011/skolkovo-agreement-1026.html (accessed March 28, 2012).

National Research Council. 1990. Recruitment, Retention, and Utilization of Federal Scientists and Engineers. Washington, D.C.: National Academy Press.

National Research Council. 2001. Review of the U.S. Department of Defense Air, Space, and Supporting Information Systems Science and Technology Program. Washington, D.C.: National Academy Press.

National Research Council. 2005. Assessment of Department of Defense Basic Research. Washington, D.C.: The National Academies Press.

National Research Council. 2009a. Beyond "Fortress America": The National Security Controls on Science and Technology in a Globalized World. Washington, D.C.: The National Academies Press.

National Research Council. 2009b. Experimentation and Rapid Prototyping in Support of Counterterrorism. Washington, D.C.: The National Academies Press.

National Research Council. 2010. The Use of Title 42 Authority at the U.S. Environmental Protection Agency. Washington, D.C.: The National Academies Press.

National Science Board. 2012. Science and Engineering Indicators 2012. Arlington Va.: National Science Foundation.

National Science Foundation. 2010. Characteristics of Recent Science and Engineering Graduates: 2006. Available at http://www.nsf.gov/statistics/nsf10318/pdf/nsf10318.pdf (accessed March 27, 2012).

OCED (Organisation for Economic Cooperation and Development). 2011. Lessons from PISA for the United States, Strong Performers and Successful Reformers in Education. Paris: OECD Publishing.

PCAST. 2010. Report to the President on Accelerating the Pace of Change in Energy Technologies Through an Integrated Federal Energy Policy. Washington, D.C.

POGO (Project on Government Oversight). 2011. Bad Business: Billions of Taxpayer Dollars Wasted on Hiring Contractors. Available at http://www.pogo.org/pogo-files/reports/contract-oversight/bad-business/co-gp-20110913.html (accessed September 13, 2011).

Roach, M., and H. Sauermann. 2010. A taste for science? PhD scientist's academic orientation and self-selection into research careers in industry. Research Policy 39(3):12.

Ryoo, J., and S. Rosen. 2004. The engineering labor market. Journal of Political Economy, 112(S1):40.

Salzman, H. 2012 (unpublished). The New STEM Labor Market Segmentation: Implications for Meeting Workforce Needs of DoD and the Industrial Base. Presentation to the Workshop on STEM Workforce Needs for the U.S. Department of Defense and the U.S. Defense Industrial Base.

Salzman, H., and B.L. Lowell. 2008. Making the grade. Nature 453:2.

Sauermann, H., and M. Roach. 2012. Science PhD career preferences: Levels, changes, and advisor encouragement. PLoS ONE 7(5):e36307.

Secretary of the Navy. 2006. Department of the Navy Information Security Program.

Sharma, Y. 2011. ASIA: "Brain reclaim" as talent returns from West. University World News. Available at http://www.universityworldnews.com/article.php?story=20110415201701401 (accessed October 3, 2012).

Stephan, P. 2012. How Economics Shapes Science. Cambridge, Mass.: Harvard University Press.

Wadhwa, V. 2011. We need to stop America's brain drain. Washington Post, October 4, 2011.

Wadhwa, V., G. Jasso, B. Rissing, G. Gereffi, and R.B. Freeman. 2007. Intellectual Property, the Immigration Backlog, and a Reverse Brain-Drain: America's New Immigrant Entrepreneurs, Part III. Kansas City, Mo.: Ewing Marion Kauffman Foundation.

Wadhwa, V., A. Saxenian, R. Freeman, G. Gereffi, and A. Salkever. 2009. America's Loss Is the World's Gain: America's New Immigrant Entrepreneurs. Kansas City, Mo.: Ewing Marion Kauffman Foundation.

Wasem, R.E. 2012. Immigration of Foreign Nationals with Science, Technology, Engineering, and Mathematics (STEM) Degrees. Washington, D.C.: Library of Congress.

Xie, Y., and A. Killewald. 2012. Is American Science in Decline? Cambridge, Mass.: Harvard University Press.

Yehle, E. 2011. No home run yet for ARPA-E, but chief says "motivated" team's on track. Greenwire, April 7.

5

Institutional Capacity in Education and the DOD Investments Needed to Ensure an Adequate STEM Workforce

INTRODUCTION

The capacity of educational institutions to educate and maintain the science, technology, engineering, and mathematics (STEM) workforce of the Department of Defense (DOD) is addressed in this chapter, including some of the impediments these institutions are facing. As in other portions of this report, this chapter focuses on three basic priorities: quality, skills mix, and flexibility. This chapter also discusses the investments that DOD is making in its current STEM workforce related to education and offers some recommendations for focusing these investments in the future.

With respect to the priority of achieving the right skills mix, DOD should focus its education and personnel efforts on the specific skill sets critical to its ongoing and anticipated needs. With this in mind, DOD support for education activities should emphasize fields that underpin its ongoing assessment of security needs, and encourage appropriate continuing education for its STEM personnel. Initially, it must be clarified what the meaning of institutional capacity is. DOD operates a substantial kindergarten through twelfth grade (K-12) school system for its overseas employees. It has a variety of postsecondary educational institutions,[1] including the military service academies, in addition to many training institutions and research facilities, all of which contribute to educating and training the DOD STEM workforce. Beyond the infrastructure that it owns, DOD maintains many relationships at all levels of civilian educational and research institutions, public and private, in the United States and beyond, that are also part of DOD's network of institutional capacity. These include the gamut of educational institutions from K-12 through universities, along with life-long learning programs that contribute to continuous workforce refreshment.

CHALLENGES TO MEETING THE EDUCATIONAL NEEDS OF THE STEM WORKFORCE

Attrition and Time to Degree Completion

Although there are STEM workers in DOD at all educational levels, the preponderance are scientists and engineers at the bachelors' degree level and above, as discussed in Chapter 3. As of 2011, approximately 48 per-

[1]Examples of these educational institutions include the Defense Acquisition University, the Air Force Institute of Technology, the Uniformed Services University of the Health Sciences, and the Naval Postgraduate School.

cent have a bachelor's degree, close to a quarter have a master's degree, and roughly 5 percent have a doctoral degree.[2] The time required to attain this level of education must be considered when projecting how readily DOD can expect to grow, or renew, its cadre of STEM employees.

In the U.S. school system, students who ultimately join the STEM educated workforce typically begin to diverge from the rest of the population by taking college preparatory mathematics and science in about the eighth grade. This is generally at least 8 years before baccalaureate graduation and 10 years before earning a master's degree. Further, the United States average for attainment of a PhD in science and engineering is 7 years from entrance into graduate school, implying substantial amounts of earnings foregone while in student status. Something to consider is that the reforms related to the Bologna Declaration of 1999 in Europe aim for a 5-year process (2-year master's plus 3-year PhD) (Kehm, 2006).

The prediction of future supply is complicated by attrition rates. The Beginning Postsecondary Students Longitudinal Study found that only 35 percent of eighth grade public school students subsequently enroll in an accredited 4-year college and 22 percent in a 2-year college, for a total of 57 percent entering a 2-year or 4-year institution (National Center for Education Statistics, 2010). Among postsecondary students entering 4-year colleges in 2003-2004, 24 percent had obtained degrees or persisted (i.e., were still enrolled) in STEM fields as of 2009 (National Science Board, 2012; Table 2-8). These same data underscore, however, that not all college freshmen who declare STEM majors graduate with such degrees. Among students entering 4-year colleges in 2004 and subsequently declaring STEM majors, the longitudinal study found roughly 80 percent still enrolled or having attained degrees (bachelor's, associate's, or certificates) in STEM fields as of 2009 (Figure 5-1). These figures are substantially lower in some STEM fields such as engineering.

Next considering all postsecondary institutions (and widening the scope to include 2-year and less-than-2-year colleges), the data show similarities in the percentage of entrants in STEM fields (Figure 5-2). The number who persisted in STEM fields or attained degrees 6 years later was, however, smaller than that for 4-year colleges alone, with 56 percent having received a degree and 14 percent still enrolled.

Visa Issues

As discussed in Chapter 4, persons who initially enter the United States on temporary visas (work or study) related to STEM fields can become citizens and thus become eligible for employment in national security related activities. International students can thus become a source of STEM hires for DOD. Further, if DOD were to adopt more liberal practices toward hiring of non-U.S. citizens as recommended in Chapter 4, the education of international students in U.S. universities would become a more important component of the STEM pipeline. This section reviews the current picture of international students in the United States and discusses other countries in which the fraction of such students is on the increase.

The United States plays host to the largest number of international students in STEM fields (Figure 5-3). China and other countries in developing Asia (the "Asia-8") are among the largest source countries of international STEM graduates in the United States (Figure 5-4). The number of full-time graduate students in science, engineering, and health fields—largely those with F-1 non-immigrant visas—was nearly 149,000 in 2009, up considerably from just over 91,000 in 1990 (Wassen, 2012). At the bachelor's degree level, temporary residents in 2009 were awarded only about 3 to 4 percent of degrees in STEM majors, although by specific major the fraction earned by such persons can be higher (e.g., electrical and industrial engineering degrees are each 9 percent).[3] Additionally, in 2009, the foreign student population earned 26.6 percent of doctoral degrees in science and 57.4 percent of the doctoral degrees in engineering (National Science Board, 2012; Appendix Table 2-19). Upon completion of their degree, foreign students on F-1 visas have various routes by which to change their immigration status to remain in the United States and become employed in STEM fields, including through acquisition of H1-B visas (Government Accountability Office, 2007), discussed in the Chapter 4 section "Security Clearances." Certain organizations such as Sandia National Laboratories have developed pathways by which a foreign national may become a member of

[2]Of the 22 percent that have attained less than a bachelor's degree, half are in the major occupational group, computer and mathematical sciences. See Chapter 3 for detailed discussion

[3]See, for example, Appendix Table 2-19 in National Science Board (2012).

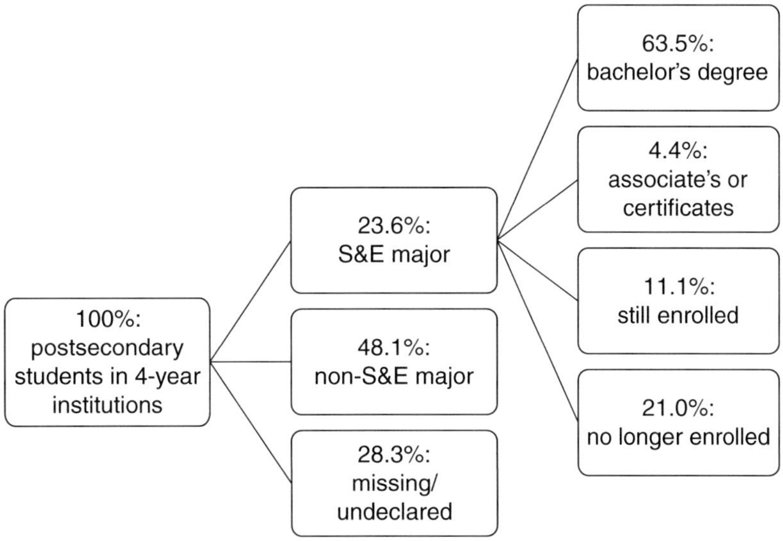

FIGURE 5-1 Persistence in science and engineering STEM fields and attainment of STEM degrees among postsecondary students in 4-year postsecondary institutions.
NOTE: Data were as of the end of the 2008-2009 academic year for the cohort that began postsecondary education in the 2003-2004 academic year. STEM fields include physical sciences and biological/agricultural sciences; engineering/engineering technologies; and computer/information sciences.
SOURCE: Department of Education (2012).

its technical staff while concurrently obtaining citizenship and security clearances (see discussion in the Chapter 4 section "Temporary Work Visas").

Several other nations, including several Commonwealth countries, have facilitated the issuing of visas to foreign nationals who earn graduate degrees in science and engineering fields in Commonwealth countries.

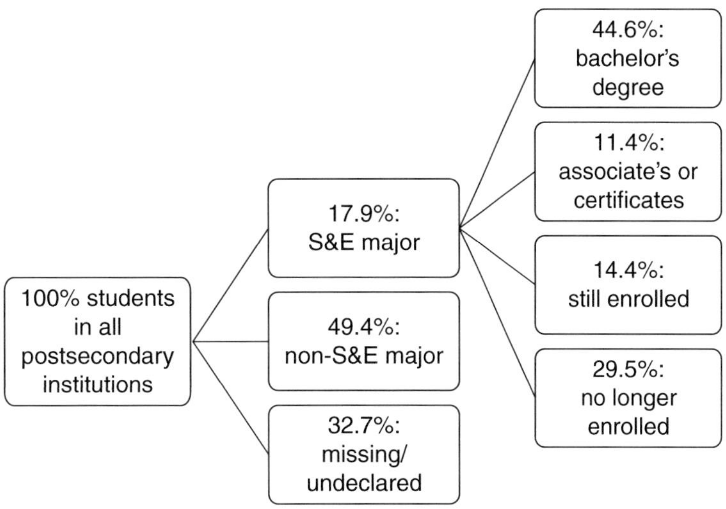

FIGURE 5-2 Persistence in science and engineering STEM fields and attainment of STEM degrees among postsecondary students entering 2- and 4-year postsecondary institutions.
NOTE: Data were as of the end of the 2008-2009 academic year for the cohort that began postsecondary education in the 2003-2004 academic year. STEM includes physical sciences and biological/agricultural sciences; engineering/engineering technologies; and computer/information sciences.
SOURCE: Department of Education (2012).

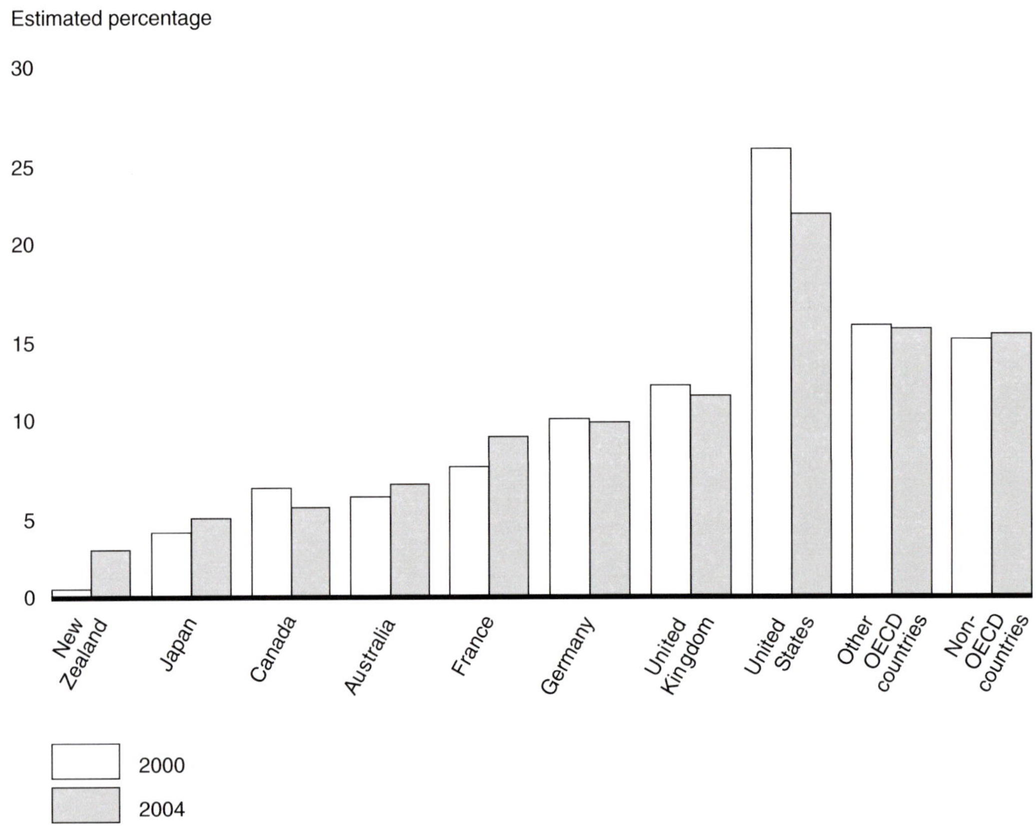

FIGURE 5-3 Estimated percentages of all international higher education students in STEM fields in a selection of countries, by country of enrollment, 2000 and 2004.
SOURCE: GAO (2007).

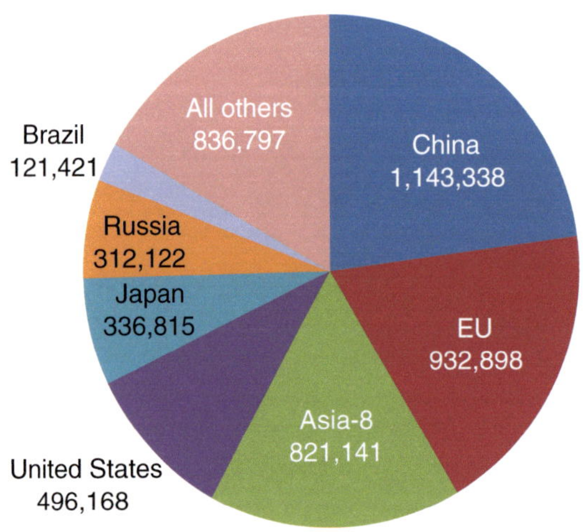

FIGURE 5-4 First university degrees in S&E fields, 2008 or most recent year.
NOTE: Asia-8 includes India, Malaysia, Philippines, Singapore, South Korea, and Taiwan; data on Indonesia and Thailand are not available.
SOURCE: Adapted from Appendix Table 2-32 in National Science Board (2012).

Immigration advocates and some observers in higher education note that close U.S. allies, including some Western European countries and Australia, Canada, and New Zealand, have substantially increased the number of foreign students studying in their universities, echoing the trend noted above in the United States.[4] In Australia, now one of the leading countries in terms of percentages of international students, 56 percent of the international students in 2009 were undergraduates. Management and commerce was the most popular field and accounted for 48 percent of the students (Phillimore and Koshy, 2010). Engineering and related technology fields, though the second most popular fields, represented much smaller percentages, only about 8 percent. One of the principal reasons for the expansion of international students in Australian universities is the government's urging that Australian universities seek increased revenues from foreign students instead of from the Australian government.

Education Costs

The U.S. Government Accountability Office found that average annual costs for private colleges and universities doubled between 1990 and 2004, from $13,237 to $26,489. However, the costs at 4-year public institutions increased by approximately 118 percent over the same time period (whereas for 4-year private colleges it was a 100 percent increase and at 2-year institutions an 83 percent increase) (Government Accountability Office, 2007). Most published fees at public institutions reflect in-state costs, and costs for out-of-state students are substantially higher. Even at 2-year institutions, the rate of tuition growth has been 3.8 percent above the rate of general inflation over the decade 2001-2002 to 2011-2012 (College Board Advocacy and Policy Center, 2011). Still, community colleges are considerably less expensive than 4-year institutions, even public ones. These rising costs of higher education have led to an increase in student loan debt, which in turn is impacting students' career choices. A 2007 report found that debt "leads graduates to choose higher-salary jobs" and that "debt appears to reduce the probability that students choose low-paid 'public interest' jobs"(Rothstein and Rouse, 2007). Some federal agencies and other organizations have student loan debt forgiveness programs that can heavily influence the attractiveness of potential employers. This coupled with any hiring delays due to security clearances may well discourage students from pursuing many DOD opportunities.

WHERE AND HOW SHOULD DOD INVEST IN EDUCATIONAL CAPACITY?

Although as discussed in Chapter 3 there is no evidence of a shortfall in the availability of a STEM workforce within DOD, except in selected disciplines, it is nevertheless prudent to consider how DOD can ensure that its STEM workforce remains robust in an increasingly uncertain future. This section first describes the current level and focus of DOD support for STEM and then discusses some specific approaches that, if adopted, could enhance the pipeline of STEM workers available to DOD.

DOD's Investments in STEM Development

Turning first to the resources DOD has to invest in this endeavor, the committee notes that these include money, people, facilities, and programs (e.g., procurement, scholarships). Table 5-1 shows how DOD's Office of the Assistant Secretary of Defense for Research and Engineering (ASD(R&E)) is currently investing in STEM development programs, although this clearly does not represent the total investment in STEM across DOD. The latter include programs in support of basic research (so-called 6.1), which alone was $2.1 billion in FY 2011 to all DOD components, and applied research and other categories of funding aimed at moving a system to a higher level of technological readiness.

However, as DOD becomes a diminishing fraction of the global demand for skilled workers, it must become more strategic and nimble in its STEM investments. The report on the federal STEM education portfolio by the Committee on Science, Technology, Engineering and Mathematics Education (co-STEM) of the National Sci-

[4]It is important to recognize significant differences, such as the fact that in Europe many such foreign students are from other European countries that are members of the EU, of NATO, or both.

TABLE 5-1 ASD(R&E) Investments in STEM

STEM Programs	FY11 Presidential Budget Request / FY11 Enacted	Targeted Group
National Defense Education Program (NDEP) K-12 Informal Education	$18M / $11.2M	K-12
Awards to Stimulate and Support Undergraduate Research Experiences (ASSURE)	$4.5M / $4.5M	Undergraduates
Science, Mathematics and Research for Transformation (SMART) Program	$56.0M / $48.8M	Undergraduates
HBCU / MI Program*	$15M / $17.3M	Faculty, staff, and students of minority institutions
National Defense Science and Engineering Graduate (NDSEG) Fellowship Program	$38.3M / $38.3M	PhD students at/near the beginning of their graduate study
National Security Science and Engineering Faculty Fellowship (NSSEFF)	$36.12M / 30.721M	University faculty, staff scientists, and engineers of accredited, U.S. doctoral degree-granting academic institutions
Presidential Early Career Awards for Scientists and Engineers (PECASE)	Army: $5.1M Navy: $5.1M Air Force: $4.5M Total: $14.7M (Enacted)	Outstanding scientists and engineers beginning their independent careers

*Funding includes monies for scholarships/fellowships and for research.

SOURCE: Laura Adolfie, DOS STEM Development Office, personal communication, December 15, 2011. Description of ASSURE based on Air Force Research Laboratory (2012); HBCU /MI based on Defense Technical Information Center (undated); NDSEG based on Office of Naval Research (undated). NDSEG based on National Defense Education Program (undated); and PECASE based on National Science Foundation (undated).

ence and Technology Council (2011) appropriately stated, "Our analysis indicates that the critical issue related to federal investments on STEM education is not whether the total number of investments is too large or whether today's programs are overly redundant with one another. Rather, the primary issue is how to strategically focus the limited federal dollars available so they will have a more significant impact in areas of national priority." In developing a strategic plan to meet DOD STEM workforce needs, it must minimize duplication of other federal or private programs, while emphasizing programs that have the greatest leverage in meeting DOD requirements. This is consistent with the findings from the Government Accountability Office (2012) report 12-108 on STEM, which led to the review and recommendations from the National Science and Technology Council report of February 2012 entitled *Coordinating Federal Science, Technology, Engineering, and Mathematics (STEM) Education Investments: Progress Report.*

This challenge of focusing limited federal dollars to improve the supply of STEM workers can be seen as having both a direct side, including investments in education, and an indirect side. Both sides must be weighed in building and maintaining an appropriate investment strategy. For example, DOD could sponsor large numbers of scholarships that carry obligations for employment in a DOD laboratory, which consequently might directly enhance its supply of educated STEM workers. The return on investment is, however, difficult to judge. For instance, should DOD invest in K-12 teacher education in STEM disciplines, or in internships for students? And if the latter, at what levels of education is intervention most strategic?

On the indirect side, DOD could fund compelling high-technology programs and laboratories that would attract high-quality STEM workers, thereby stimulating its workforce through contractors. The DOD investment priorities for programs and facilities would cascade from national security priorities. Here the investment will be most effective if it can avoid year-to-year fluctuations, thereby providing a stable signal to would-be researchers (JASON, 2009); providing $100 million evenly over 10 years, for example, is preferable to a short burst of very high funding (National Research Council, 2012b). The Defense Science Board has further underscored the importance of stable funding and has explicitly recommended that DOD take steps to eliminate large fluctuations in funding with its 6.1 (i.e., basic research) programs (Defense Science Board, 2012).

There are also the matters of stimulating currently underrepresented populations and tapping more heavily into the global pool of STEM workers. This report focuses primarily on the DOD civilian STEM workforce, stipulating that the military services also have need for these skills, but acknowledging that responsibility for recruiting, training, and equipping uniformed forces are in the portfolios of the individual services.

Investing Across the Education Continuum

K-12 Intervention

As discussed earlier, the DOD maintains a system of K-12 schools, primarily overseas, that caters to dependents of DOD employees. These are typically well run and could be used in to address the current issue in at least two ways. The schools can provide examples of best practices to state and local schools, and they can be used as workshops for experimental approaches to pedagogy and education administration.

Improving K-12 STEM education on a nationwide basis is beyond the role of DOD; K-12 education is heavily controlled and financed at state and local levels. DOD does not have the mandate, resources or know-how to engage the U.S. STEM challenge as a whole. The National Defense Education Act of 1958 (Public Law 85-164) is not a realistic model for national action today. However, DOD could make a significant difference in the K-12 space by mining the talent of its own employees' families the way some companies do. DOD school-age dependents include more than 1.1 million children (Government Accountability Office, 2011), 88,000 of whom are enrolled at the DOD Education Activity (DODEA) (Department of Defense Education Activity, 2012), with most of the rest in public schools near large bases. A dedicated effort to nurture the potential of these students could involve DOD-themed STEM learning opportunities in school and participation in competitions such as FIRST Robotics and MATHCOUNTS, summer programs at DOD labs, and internships. A department-wide K-12 focus on military dependents would not preclude broader outreach for STEM talent at the postsecondary level, but it would be much more likely to produce results than the current broad approach. Some advantages of a new approach might include the following:

- *Politics.* A "grow your own" focus on DOD STEM development ties together two widely shared national priorities—STEM education and military families. It also provides a K-20 component to the Administration's efforts to encourage returning veterans to pursue careers in STEM.
- *DOD synergy.* Within DOD, the "targeting" of military dependents leverages the interests of the Assistant Secretary of Defense for Research and Engineering (ASD(R&E)), where the need lies, with those of the Office of Personnel and Readiness (P&R), which has the credibility in K-12 STEM education through DODEA. If the fragmented assets of DOD were melded together, the department could more effectively impact the K-12 level and build a stronger base for its projected workforce needs.
- *Diversity.* The military dependent pool is very diverse. DOD's diversity initiatives have varied greatly at the K-12 level. The proposed strategy would bring more focus and help level the playing field for underrepresented minorities and other diverse groups.

Obviously, many hard questions need to be asked. Is there any empirical evidence that "growing your own" works in STEM education? Would there be opposition to giving special attention to particular pools of youth? Is DOD capable of breaking through its own stovepipes? Nonetheless, an initiative along these lines has the potential to break important new ground and thus deserves serious consideration.

Lastly, the Secretary of Education can have a significant impact through federal oversight and other tools such as applying conditions to the awarding of grants. Coordination between DOD and the Department of Education on mutual goals should be enhanced.

Competitive Internships Recruitment

If DOD were able to identify STEM-inclined students in high school and provide opportunities for them to be exposed to those sciences of importance to DOD, the latter could build the pipeline and also keep those students close to the work done by DOD, thus improving the chances of recruiting them.

Role of Community Colleges

Another possible approach to respond to the problem of ensuring that DOD has the necessary STEM talent is to strengthen community college programs so that the first 2 years of education could be provided by these less costly institutions. Community colleges have been around for about 100 years, and there are now some 1,100 public community colleges around the country. There are an additional 200 or so private 2-year institutions. These colleges are, for the most part, "open-door" institutions that allow enrollment by students who do not have high academic scores or preparation. Most students are given an entrance test, designed to evaluate their readiness for college-level work. They are then placed in courses appropriate to their skill level. Outside the community college sector this screening process is not well known, and as a result, community colleges often suffer from the stigma of being institutions of lower standards and, by extension, lesser quality. Yet, those students who transfer to 4-year institutions from community colleges perform quite creditably, graduating at rates comparable to, if slightly below, those of 4-year students and with the same or higher GPA.[5]

Another approach to the pipeline issue would be to link community colleges, high schools, and neighboring universities into alliances that would identify students with demonstrated math and science aptitude as early as tenth grade and create pipelines from that point on through the community college and into the 4-year college or university to complete the baccalaureate degree. There are a number of pieces that would need to be put in place, some through the auspices of DOD. Relationships that could be facilitated by DOD would have to be built among the participating institutions, and joint institutional expectations would have to be developed. Of critical importance are the issues of standards and quality, and these can be assured only by teachers and faculty working together to build curricula and to develop appropriate pedagogy.

Some community colleges have already developed structures called dual-enrollment or concurrent enrollment that allow students to take courses at a community college for college credit while still in high school. In some of these arrangements, these courses also count as fulfilling high school requirements, thus accelerating the time to degree completion. Similarly, large numbers of students already transfer from 2-year to 4-year colleges. The proposed model would streamline these existing models in a seamless process that monitors student progress from tenth grade to completion of the baccalaureate degree.

There is a concern that community colleges have moved away from the technical mission, which was their focus prior to the 1980s. Unfortunately, there has been little empirical research exploring this very important trend. Upon consultation with a series of researchers from the American Association of Community Colleges, the committee found that interest in technical-vocational programs decreased drastically due to the decline in such programs at the high school level. This was particularly evident during the 1980s in urban school districts where African American parents, worried about the bifurcation of the workforce and collegiate preparation going to whites and vocational education going to blacks, insisted that their children be given the same academic education as other populations. The researchers also postulated that the decline in the industrial base of the United States over the last quarter of a century caused the decline in the community colleges' technical mission. As many of the blue-collar jobs were off-shored, it became less urgent for institutions to provide the skills for what were essentially dying industries.This also led to the up-skilling of the workforce. Whereas high schools were able to provide the training for many of the disciplines needed in the workforce for technical jobs, the current skill level demands are well beyond the high school institutional capacity. These skill requirements are elevated to higher education, particularly the community colleges.

However, the researchers also stated that although the technical mission appears to be on the decline, the

[5]See, for example, pp. 33-34 of National Academy of Engineering and National Research Council (2005).

numbers both of certificate program completions and of certificates awarded in the last 20 years have actually increased. Since certificates are almost never credit transfer-related, but rather have workforce applicability, this increase might be taken as evidence that the curriculum has become more workforce-driven in the last two decades. However, the committee has no clear indication that these are technical certificates, and in fact believes that they may be more likely to be awarded in business-related areas of study. More research on the current condition of technical programs is required in order to have a full understanding of the state of community colleges and their potential role in the development of the DOD STEM workforce.

Bachelor's, Master's, and Doctoral Level

Over 5,000 undergraduate and graduate students are provided support by DOD through research assistants, research awards, and other mechanisms such as the NDSEG (see Table 5-1) (Defense Science Board, 2012). Among these latter mechanisms, DOD currently has a scholarship-for-service program for civilians similar to the Reserve Officers' Training Corps (ROTC): the Science, Mathematics and Research for Transformation Scholarship for Service Program (SMART) enables students pursuing an undergraduate or graduate degree in STEM disciplines to receive a full scholarship and be gainfully employed upon degree completion. Participants in SMART have the opportunity to pursue summer internships at DOD laboratories, giving them exposure to a research environment and encouragement to pursue a career in STEM. The program seeks students pursuing degrees in aeronautical and astronautical engineering, biosciences, chemical engineering, chemistry, civil engineering, cognitive, neural, and behavioral sciences, computer and computational sciences, electrical engineering, geosciences, industrial and systems engineering, information sciences, materials science and engineering, mathematics, mechanical engineering, naval architecture, ocean engineering, nuclear engineering, oceanography, operations research, and physics. Upon selection, awardees are assigned to the DOD laboratory where he/she is expected to serve as a paid summer intern and complete a 1-year for 1-year period of post-graduation employment as a DOD civilian. This SMART program can be managed to adjust the input as DOD's needs evolve.

At the PhD level, one of the best sources of quality talent will be the doctoral research assistants supported by DOD research grants to universities for basic and applied research.

Postdoctoral Level

The DOD will need to be able to recruit scientists and engineers at the postdoctoral level whose expertise may be multi-disciplinary and who may be eager to have support for their research as well as access to equipment and laboratories. As noted in the interim report for this project, "The DoD, as viewed by the top STEM talent pool, must become an attractive career destination for a suitable share of the most capable scientists, engineers, and technicians" (National Research Council, 2012a). The workshop on DOD STEM workforce needs, convened in August 2011 as part of the present study, included discussion of the possibility that "DoD could offer fellowships aimed at bringing people to its laboratories and immersing them in DoD problems" (National Research Council, 2012b). It was noted that such fellowships would need to be of a caliber that could compete with other well-regarded fellowship programs such as those of the National Science Foundation or the National Institutes of Health.

DOD provided the funding for 2,034 postdoctoral fellowships on tenure in 2010, compared to the federal government-wide total of 24,367 such positions, of which roughly half were sourced to the Department of Health and Human Services (National Science Foundation and National Institutes of Health, 2010). DOD can in addition host postdoctoral associates funded by other sources, for example, the Department of Homeland Security and the National Institutes of Health.

The ASD(R&E) funds postdoctoral associates, for example, through the National Security Science and Engineering Faculty Fellowship (NSSEFF; see Table 5-1), which "engages approximately 250 of their researchers, students and postdocs in the DoD and research challenges" (Lemnios, 2011). The three military services offer highly competitive postdoctoral research opportunities, which can often lead to participants becoming involved in further DOD activities (Defense Science Board, 2012). For example, the Naval Research Laboratory currently hosts just over 100 fellows: 80 through the National Research Council's Research Associateship Program, 20 through the American Society for Engineering Education, and a handful through other sources.

A report by the JASONs recommended expanding DOD's postdoctoral fellowships in line with a new business model of vertical integration at the level of undergraduate students through faculty so as to assist DOD in reinforcing its external network of researchers as well as its "brand" (JASON, 2009, p. 54). The Defense Science Board (DSB) recommended that the number of scientists and engineers in postdoctoral positions in DOD laboratories be greatly expanded. The DSB further noted that DOD's NSSEFF program had not recruited a new class in over a year as of December 2011 (Defense Science Board, 2012).

Professional Science Master's Degree

Professional Science Master's (PSM) degree programs have been expanding around the country and now number over 250 at some 117 institutions. These programs have created a distinctive approach to articulating curricular design, with scientific/engineering workforce needs specified by employers (Professional Science Master's, 2012; National Research Council, 2008). Often these programs are cross-disciplinary, and new programs of this type could be configured to meet the broad skills specified as needed by DOD management. PSM curricula are created by faculty, but in direct consultation with employer advisory committees that continue to advise as needs and programs evolve. In addition, some PSM programs (such as those at California State University) have been actively seeking ways to articulate with community colleges.

So far, most employers involved in PSM programs have been corporate; DOD has not been involved to any significant degree. However, if DOD agencies could describe PSM programs that would meet their projected needs, possibly in concert with large procurement programs, and commit to offer summer internships to students and to hire recent graduates with such capabilities, PSM degrees would likely be configured to meet DOD's needs by a number of universities that are actively expanding their PSM offerings. As noted, these degrees could also be designed to articulate with community colleges. If DOD could offer PSM students even partial financial support—in return for appropriate commitments for specified years of DOD service—this could be a powerful recruitment mechanism. Moreover, students identified as promising future hires by DOD agencies could be pre-cleared while still in student status and then be ready to begin productive careers with DOD with no delay when they graduate.

Lifelong Learning Continuum

The DOD has many workforce development and executive education programs that are targeted at various facets of the STEM workforce. While most of these programs are concentrated on uniformed service members, there is vast potential for expansion of the programs to the civilian DOD workforce. These run the gamut from short courses taken in the work place on-line to advanced, graduate level in-residence courses. Some DOD institutions offer certificate and degree credit courses in a broad array of resident, non-resident, and hybrid programs. The Defense Acquisition University, for example, provides training at three levels of certification. These DOD programs are obviously important for developing, maintaining, and enhancing the relevance of the skills of DOD workers.

There are also DOD institutions that can be exploited to convert non-STEM employees into STEM workers. For example, the Naval Postgraduate School (NPS) has a program to re-qualify mid-career naval officers with non-STEM degrees into master's degree graduates of science and engineering programs. It does so in both resident and non-resident offerings, with effort focused in the interest of economy of time and money.

DOD civilians are eligible for these NPS programs and many are enrolled now, especially in the systems engineering domain, but further use could be made in the future of this conversion option, short-circuiting as it does the 8 to 10 year lag time between eighth grade and the workplace, and virtually eliminating attrition and clearance issues.

Other technology-focused DOD institutions that are accredited at the graduate degree level can be similarly exploited, including the Air Force Institute of Technology (AFIT), the Information Resources Management College

(IRMC), and perhaps others. Other high-quality online courses and materials are increasingly available, such as the MIT/Harvard "edX"[6] initiative, and could be used effectively by the DOD.

FINDINGS AND RECOMMENDATIONS

Finding 5-1. DOD's external funds for STEM education are limited. The vast majority of K-12 education is controlled and financed at the state and local level.

Recommendation 5-1a. The DOD should be more strategically focused to maximize its leverage on STEM workforce issues that have high priority for DOD. At K-12 levels, the focus of DOD's funds should be on DOD schools; those public and charter schools in locations with a large DOD presence; and outreach to other schools by DOD STEM personnel.

Recommendation 5-1b. In higher education, DOD should deploy its limited resources to:

- Continue and expand support for DOD's SMART program with particular emphasis on DOD area needs.
- Sponsor U.S. universities to develop Professional Science Master's (PSM) degrees configured for DOD workforce needs, including those discussed in Chapter 2 (e.g., information technology, including cyber security and cyber warfare; autonomous systems; systems biology; innovative materials; and efficient manufacturing). DOD agencies could help university faculty plan such degrees; offer PSM students internships and hire PSM graduates; and provide partial financial support to PSM students in return for appropriate DOD service.

Recommendations 5-1c. The DOD should encourage and support alliances to more effectively link high schools, community colleges, and universities in identifying and encouraging secondary school students who have demonstrated aptitude in math and science but who, for family or economic reasons, may not be planning on entering directly into 4-year institutions. Such alliances might prove especially effective in attracting underrepresented minority students.

Finding 5-2. The DOD supports programs and in-house institutions that make STEM education, across a wide range of disciplines, available to its workforce at all levels of education. Some of these programs require in-residence training/education, although many deliver curricula remotely, including to the workplace. One particularly successful example is the Naval Postgraduate School's program for rapidly reeducating personnel with no STEM educational background to a master's degree standard in technical fields. The DOD could also benefit from certificate and master's degree programs, created jointly with universities and targeted specifically to DOD workforce needs for advanced education. These topical programs could be delivered on-site and/or on-line as best serve DOD circumstances.

Recommendation 5-2. Because DOD's STEM needs evolve, a strategic assessment of DOD's own STEM training/education capacities should be undertaken periodically to ensure that its capabilities to prepare its existing workforce to serve DOD needs is sufficient. As a follow up to this assessment, DOD should create/adapt programs in support of its STEM professionals to maximize their currency in this rapidly changing science, technology, and DOD program/project management environment. The DOD effort could also include creating certificate and professional master's degree programs developed in partnership with universities and possibly industry, whose content specifically targets the educational and skills needs identified by DOD.

Finding 5-3. The availability of stable funding for basic and applied research is an important factor in building and maintaining a robust STEM workforce. Also, it is apparent that those who become DOD STEM workers are drawn from a wide range of colleges and universities.

[6]For more information, see Massachusetts Institute of Technology (MIT) News (2012).

Recommendation 5-3. The DOD should maintain its commitment to stable basic research funding across a broad spectrum of U.S. colleges and universities in STEM areas of importance.

Finding 5-4. Integration of postdoctoral fellows into the DOD STEM mission is the fastest, most cost efficient way to recruit and screen PhDs for future career employment while making them aware of exciting DOD opportunities. Postdoctoral fellowships have been largely ignored in favor of higher-cost support of graduate students whose expertise (selected 6 years in advance) may not align with the rapidly changing needs of DOD. Although DOD has contracts to pay postdoctoral fellows through the National Research Council and the American Society for Engineering Education, among others, the funds come directly from laboratory operating budgets and compete in many cases with funds for staff salaries. A DOD-wide postdoctoral fellowship program that covers all costs of the fellow to the laboratories would be most cost-effective.

Recommendation 5-4. The DOD should initiate a postdoctoral fellowship program for recruitment of the highest-quality STEM graduates into the DOD laboratories that covers all costs of the fellowships. The applications should include inputs from both the postdoctoral candidate and the doctoral research mentor.

REFERENCES

Air Force Research Laboratory. 2012. AFOSR: Awards to Stimulate and Support Undergraduate Research Experiences (ASSURE). Available at http://www.wpafb.af.mil/library/factsheets/factsheet.asp?id=9333 (accessed June 6, 2012).

College Board Advocacy and Policy Center. 2011. Trends in College Pricing. Washington, D.C.: College Board.

Defense Science Board. Undated. Report of the Defense Science Board Task Force on Basic Research. Available at http://www.acq.osd.mil/dsb/reports/BasicResearch.pdf (accessed April 4, 2012).

Defense Technical Information Center. 2012. Historically Black Colleges and Universities (HBCU) Program. Available at http://www.dtic.mil/dtic/aboutus/dodprograms/hbcu.html (accessed April 4, 2012).

Department of Defense Education Activity. 2012. Enrollment--Present. Available at http://www.dodea.edu/datacenter/enrollment_display.cfm (accessed March 23, 2012).

Department of Education. 2012. 2003-04 Beginning Postsecondary Students Longitudinal Study, Second Follow-Up (BPS:04/09). National Center for Education Statistics.

Government Accountability Office. 2007. Higher Education: Challenges in Attracting International Students to the United States and Implications for Global Competitiveness. Washington, D.C.: Government Accountability Office.

Government Accountability Office. 2011. Education of Military Dependent Students. Washington, D.C.: Government Accountability Office.

Government Accountability Office. 2012. Science, Technology, Engineering, and Mathematics Education: Strategic Planning Needed to Better Manage Overlapping Programs across Multiple Agencies. Washington, D.C.: Government Accountability Office.

JASON. 2009. S&T for National Security. McLean, Va.: MITRE Corporation.

Kehm, B.M. 2006. Doctoral Education in Europe and North America: A Comparative Analysis, The Formative Years of Scholars. London: Portland Press Ltd.

Lemnios, Z.J. 2011. STEM Workforce Development for the Department of Defense. Presentation to the Workshop on STEM Workforce Needs for the U.S. DOD and the U.S. Defense Industrial Base, Rosslyn, Va., August 1.

MIT News. 2012. MIT and Harvard Announce edX. Available at web.mit.edu/newsoffice/2012/mit-harvard-edx-announcement-050212.html (accessed September 20, 2012).

National Academy of Engineering and National Research Council. 2005. Enhancing the Community College Pathway to Engineering Careers. Washington, D.C.: The National Academies Press.

National Center for Education Statistics. 2010. Graduation rates of previous year's 12th-graders and college attendance rates of those who graduated, by selected high school characteristics: 1999-2000, 2003-04, and 2007-08. Digest of Education Statistics. Table 210. Available at www.nces.ed.gov/programs/digest/dll/tables/dt10_210.asp.

National Defense Education Program. undated. Programs. Available at http://www.ndep.us/ProgNSSEFF.aspx (accessed September 19, 2012).

National Research Council. 2008. Science Professionals: Master's Education for a Competitive World. Washington, D.C.: The National Academies Press.

National Research Council. 2012a. An Interim Report on Assuring DOD a Strong Science, Technology, Engineering, and Mathematics (STEM) Workforce. Washington, D.C.: The National Academies Press.

National Research Council. 2012b. Report of a Workshop on Science, Technology, Engineering, and Mathematics (STEM) Workforce Needs for the U.S. Department of Defense and the U.S. Defense Industrial Base. Washington, D.C.: The National Academies Press.

National Science and Technology Council. 2011. The Federal Science, Technology, Engineering, and Mathematics (STEM) Education Portfolio. Washington, D.C.: Executive Office of the President.

National Science and Technology Council. 2012. Coordinating Federal Science, Technology, Engineering, and Mathematics (STEM) Education Investments: Progress Report.

National Science Board. 2012. Science and Engineering Indicators 2012. Arlington, Va.: National Science Foundation.

National Science Foundation. undated. Presidential Early Career Awards for Scientists and Engineers. Available at http://www.nsf.gov/od/oia/activities/pecase/ (accessed September 19, 2012).

National Science Foundation and National Institutes of Health. 2010. Survey of Graduate Students and Postdoctorates in Science and Engineering. Available at http://www.nsf.gov/statistics/srvygradpostdoc/ (accessed March 3, 2012).

Office of Naval Research. undated. DoD National Defense Science & Engineering Graduate Fellowship Program. Available at http://www.onr.navy.mil/Education-Outreach/undergraduate-graduate/NDSEG-graduate-fellowship.aspx (accessed September 19, 2012).

Phillimore, J., and P. Koshy. 2010. The Economic Implications of Fewer International Higher Education Students in Australia. Perth, Australia: John Curtin Institute of Public Policy.

Professional Science Master's. 2012. Available at www.sciencemasters.com (accessed March 27, 2012).

Rothstein, J., and J.C.E. Rouse. 2007. Constrained After College: Student Loans and Early Career Occupational Choices. Cambridge, Mass.: National Bureau of Economic Research.

Wassen, R.E. 2012. Immigration of Foreign Nationals with Science, Technology, Engineering and Mathematics (STEM) Degrees. Washington, D.C.: Congressional Research Service.

6

A Closing Perspective on the DOD Workforce

The data made available to this study indicate that for the foreseeable future the pool of available STEM professionals in the United States is large enough to support the Department of Defense's (DOD's) needs, except possibly in a few select areas like certified cybersecurity professionals (Brannen and Fryer-Biggs, 2012; Center for Strategic and International Studies, 2010). This assertion is made because DOD's needs represent only a small fraction of the total nationwide requirement. The nation as a whole may face a numerical shortage of STEM workers now or in the future, but the committee has little evidence that this issue will directly impact DOD's ability to meet its personnel needs. As noted in Chapter 3 (in the section "STEM Workforce in the Defense Industrial Base"), and as the committee learned at its August 2011 workshop, there is evidence that DOD and its industrial base do not always have the right number of the right skills and the right quality at the right place and time (Swallow, 2011). However, this issue seems to be less a strategic numerical issue and more a problem of effective management within the department and organizations that impact DOD. It is the committee's assessment that the core of DOD's problem is about the management of demand, and not about supply. The fundamental issue is quality, agility, and skills mix in the DOD STEM workforce. The recommendations in this report address this issue. Less-than-effective management of the DOD's STEM workforce inhibits recruiting and retention by limiting career growth, underutilizing employee skills, and constraining the available pool of talent.

The committee notes that the attention paid to the education and career development of uniformed personnel is much greater and more disciplined than that paid to civilian DOD employees. It believes that more attention to civilian career development would benefit both DOD and its employees, especially given the challenges that lie ahead.

The evolution to a globalized economy and a globalized industrial base has put substantial management stress on the way DOD conducts its mission. The overseas migration of manufacturing and R&D activities relevant to defense (such as batteries and semiconductors) not only poses acquisition challenges, but also requires DOD to engage more openly with non-U.S. STEM professionals. In particular, the DOD and the Congress have not resolved their competing desires to acquire capable and cost-effective systems and services from a global base, while supporting the U.S. economy and buying from trusted U.S. suppliers. One result of this issue is an acquisition process that puts under-empowered personnel into the nearly impossible position of trying to resolve the dilemma at their level. There is a need for a more comprehensive set of policies that will guide DOD in its interaction with the global market place. This effect can be seen in export control policies that assume that the technologies in the United States are always the best and worth keeping locked up (Epstein, 2011; National Research Council, 2009; Vest, 2011). Increasingly, expertise and innovation in core technologies are occurring outside the United States, and finding qualified STEM professionals will require new models for hiring and engaging with universities and industry overseas.

Finding 6.1. New technological advancements, often from outside the defense sector and from abroad, are appearing at an increasing rate. Adapting to this new environment requires transformational and long-term changes within the DOD management of its STEM workforce.

Security requirements, including classification and compartmentalization, pose an additional challenge to attracting and retaining high-quality STEM professionals. Engagement with outside researchers is beneficial to the careers and ongoing education of DOD professionals, but those working on classified programs often find few such opportunities. The DOD STEM workforce, particularly those who work in highly sensitive areas, need a range of regular opportunities to interact and share with colleagues in the private sector and academia. These exchanges can be structured around unclassified aspects of DOD work, or even around general research topics of benefit to DOD.

Finding 6.2. Working in classified environments can lead to professional isolation and can have a negative impact on those experts who might otherwise benefit from greater exposure to the discourse in the broader scientific community in which innovation and technology are accelerating.

BARRIERS TO ATTRACTING AND RETAINING THE BEST AND THE BRIGHTEST

Many STEM assignments in DOD involve a degree of procedure and bureaucracy that high-quality STEM professionals are unlikely to find satisfying, particularly in comparison to the academic environment. This issue is a particular challenge in the acquisition workforce. STEM skills and education are critical to understanding and evaluating DOD systems, but the nature of day-to-day work is often focused more on program management than science and engineering. Providing meaningful opportunities for technical work while developing management skills is critical to attracting and retaining STEM professionals in the acquisition workforce.

Bureaucratic obstacles also inhibit the recruiting and hiring process. The process is impersonal, slow, and often opaque to the prospective employee. It is not "owned" by the immediate organization in which the particular position exists.

In addition, there will likely be a need to hire STEM professionals in non-traditional fields for which there is neither the current focus on hiring nor a way for their expertise to be properly compensated. Work inspired by the conflicts in Iraq and Afghanistan, for example, has highlighted the importance of sociology and anthropology. On-going investment in these fields, even as the conflicts wind down, will help DOD attract and retain a relevant workforce by indicating the need for these non-traditional skills. The committee does not, however, detect any significant level of senior management focus and actions to address these problems. These problems are serious, and they deserve attention from the department.

The data on comparable salaries for STEM professionals between DOD and the private sector are complicated by many situational factors. For example, CBO finds that individuals in the federal workforce with a professional or doctoral degree earn (in wages and benefits) about 18 percent less than their counterparts in the private sector (Congressional Budget Office, 2012). However, pay differential is but one factor to consider in recruiting and retaining a high-quality workforce. The potential to serve important national interests by taking a DOD position that is demonstrably useful to the country can be a powerful lure to prospective, highly skilled professionals. Acting to ensure that its positions have these attributes provides a comparative advantage for DOD.

The DOD has several statutory options for specializing its recruiting and retention practices but does not apply them in a comprehensive and coherent manner. A structured program of experimentation in new practices, perhaps focused around the recruitment of professionals in non-traditional STEM fields such as cybersecurity, would provide an improved understanding of the available methods. Such a program would lay the foundation for a rigorous application of specialized recruiting and retention practices to address gaps in the quantity or skills of DOD STEM professionals as they arise in the future.

Finding 6.3. The STEM issue for the DOD is the quality of its workforce, not the quantity available. The DOD needs a suitable share of the most talented STEM professionals. The decisions they make within DOD are highly leveraged, impacting the efforts of very large numbers of people and enterprises both inside and outside the government.

A PERSPECTIVE ON THE INDUSTRIAL BASE

The testimony received by the committee and all of the data collected indicate that the major industrial suppliers of DOD are doing a good job of anticipating traditional and non-traditional STEM needs and acting aggressively to ensure that they have talent available.[1] They are also doing their part in supporting activities that will improve the available talent by offering educational opportunities and career development programs as a part of their recruitment and retention process. Of course, the situation could change. Several of the recommendations made in this report will apply to the industrial base as well as to the government.

Finding 6.4. The career development support for the DOD uniformed STEM workforce is excellent, whereas the career development support for the DOD civilian STEM workforce is far less developed. The defense-related industry lies somewhere between them.

ON THE UNCERTAIN FUTURE

As always, the future facing DOD is fraught with many sources of uncertainty. However, the committee does not think that the current level of uncertainty is unprecedented.

The committee was made aware of various lists of emerging new technologies that might have potential value to DOD, including one from ASDRE.[2] While not all lists are the same, they are consistent enough to use as a basis for addressing the question of uncertainty and how to deal with it. In the committee's judgment, all of the listed technologies promise future value, but it is not clear that any of them are likely to be, by themselves, game-changers, as was the case for nuclear weapons, digital electronics, and information systems.

In the face of the uncertainties in how technology will evolve, as well as the larger questions posed by geopolitical events, there is a temptation to try to forecast the future and take significant actions now in anticipation of that future. However, the committee lacks confidence that these technology forecasts can be accurate enough to rely on as a strong basis for planning. The committee is supported in this skepticism by noting the nation's past demonstrated inability to provide accurate forecasts.[3] The committee believes that this lack of confidence in forecasting argues for a more incremental approach, as well as for personnel policies that will increase the department's flexibility in adapting to unforeseen requirements. The committee thinks that the estimated sources of uncertainty described above are very likely to provide sufficient warning to permit an adequate incremental response to technology exploitation.

WHAT TO DO?

Dealing with Uncertainty

DOD should invest in emerging technologies with levels and priorities indicated by an assessment of their potential value to DOD. These investments will advance knowledge, mature understanding, and develop expertise in new fields. As these emerging technologies prove their value and increase in importance, more money and people will flow to the fields through DOD and congressional appropriations. The situation will evolve over time, and there will not be an unforeseen need for large workforce changes. The firms in the industrial base will use

---

[1] See, for example, pp. 40-44 of National Research Council (2012).
[2] See, for example, pp. 8-18 of National Research Council (2012).
[3] See, for example, Anders (2008).

these indicators to shape their own workforce in support of their business strategies, given the signals of priorities and commitments of DOD organizations. This evolutionary approach has been the U.S. historical norm, and the committee does not see that a change in strategy is needed or appropriate.

In addition to investing in technological progress in accordance with DOD judgment on priorities, the committee suggests that DOD focus on a few difficult problems that, if solved, would make significant changes in military capabilities. By forcing a focus on important missions, there will be a need to integrate technology with engineering and related systems and operational concepts. In current military operations, a focus on mitigation of improvised explosive devices has had a significant impact on the ability to integrate solutions. The committee would estimate that "portable power" could be another example whereby a new technology could be integrated into new solutions. These hard problems should be related to enduring military needs, e.g., the need to know where the enemy is, the need to know where one's own troops are, the need for rapid maneuver, and so on.

Finding 6.5. To address enduring military needs, there is an opportunity to integrate technology with engineering and related systems and operational concepts in current and future operations. Some examples include technology for negating improvised explosive devices (IEDs); portable power; and technology for non-intrusive identification of individuals and tracking of their location.

Improving the Quality of DOD Assignments

The committee recognizes that systematically improving the quality of the assignments in DOD is a massive task that involves the whole institution and how it does business. The committee has had neither the time nor the resources to attempt to address this issue in any comprehensive way. However, this aspect of DOD's situation represents a fundamental problem that inhibits DOD's access to the best and brightest talent, and it is worthy of the attention of the most senior management levels of the department. Making DOD employment attractive to the most qualified and motivated professionals will pay enormous dividends to the department and the nation.

Clarifying the overall objectives of the department and establishing confidence that the government leadership supports these objectives are a necessary foundation. Focusing on the assignment of responsibilities, authorities, and modes of accountability, and holding to these assignments in execution, are also a necessary foundation of sound management. A process of surveying and assessing the quality of each assigned position is a necessary part of any effort to change the environment.

Each of the Armed Services has an elaborate process for attending to the education, training, and career development of its military professionals. The committee believes that a similar focus on the civilian workforce would cause STEM professionals to view a DOD career more favorably. There are a number of ways that this might be enabled, including temporary assignment in industry, academia, or overseas.

There are now several rapid response organizations in DOD[4] aimed at providing quick response to commanders' needs in current conflicts Although the impetus for this capability was the conflicts in Iraq and Afghanistan, the rapid injection of some technologies and capabilities into the field should continue. It is a domain that can give STEM professionals more diverse experience and immediate feedback than can participating in a long major system acquisition cycle.

Finding 6.6. Reliable forecasting of the STEM skills needed by the DOD beyond the near term is simply not possible because of the increasing rates of advancement in science and technology and the unpredictability of military needs. *Flexibility, capability,* and *relevance* in the DOD STEM workforce are the essential characteristics sought.

[4]See discussion in the Chapter 3 subsection "Quality of Work."

Improving the Processes for Recruitment and Retention

The committee's principal recommendation for improving recruitment and retention is that the department as a whole should prioritize the issues and demonstrate a sustained, serious focus through well-advertised and aggressive actions.

The recruiting process should be made more personal for the potential employee. He or she should have information about the assignment and the supervisory structure of the position. Someone should be assigned to shepherd the paperwork associated with the hiring process. The potential employee should receive status reports on the progress of the process.

For a high-priority set of positions and potential employees, such as cybersecurity professionals, DOD should continue to exercise the authority to temporarily employ an individual while waiting for the clearance process. This action would require the development of useful activities for this period. In addition, expanding DOD's internship program, sponsoring summer-hire programs, and identifying talent early could allow the clearance process to begin while high-potential individuals are still completing their degrees. In either case, the department should also take aggressive action to shorten the period required for completing a clearance.

DOD should continue as well as expand broadly available scholarship programs (such as SMART) that are aimed at improving the quality of its current and potential employees and are tied to a commitment to service. We believe this action would be valued by the employee and would demonstrate the priority DOD places on the employee.

REFERENCES

Anders, G. 2008. Predictions of the past. Wall Street Journal, January 28.

Brannen, K., and Z. Fryer-Biggs. 2012. U.S. short on offensive cyber experts. Defense News, July 2.

Center for Strategic and International Studies. 2010. A Human Capital Crisis in Cybersecurity. Washington, D.C.: Center for Strategic and International Studies.

Congressional Budget Office. 2012. Comparing the Compensation of Federal and Private-Sector Employees. Washington, D.C.: Congress of the United States.

Epstein, G.L. 2011. The National Security Imperative for Global S&T Engagement. Presentation to the Committee on STEM Workforce Needs of the U.S. DOD and the U.S. Defense Industrial Base, Washington, D.C., September 19.

National Research Council. 2009. Beyond "Fortress America": National Security Controls on Science and Technology in a Globalized World. Washington, D.C.: The National Academies Press.

National Research Council. 2012. Report of a Workshop on Science, Technology, Engineering, and Mathematics (STEM) Workforce Needs for the U.S. Department of Defense and the U.S. Defense Industrial Base. Washington, D.C.: The National Academies Press.

Swallow, E. 2011. Northrop Grumman Corporation. Presentation to the Workshop on STEM Workforce Needs for the U.S. Department of Defense and the U.S. Defense Industrial Base, Rosslyn, Va., August 1.

Vest, C.M. 2011. STEM Workforce Needs for U.S. DOD and Defense Industry Base. Presentation to the Workshop on STEM Workforce Needs for the U.S. Department of Defense and the U.S. Defense Industrial Base, Rosslyn, Va., August 1.

Appendixes

Appendix A

Committee Biographies

Norman R. Augustine (NAS/NAE), *Co-chair,* is the retired chairman and chief executive officer of the Lockheed Martin Corporation, the nation's largest defense contractor, and a former under secretary of the Army. He is often compared to Microsoft chairman Bill Gates and former Intel CEO Craig Barrett for his national leadership in technology. He is a longtime proponent for ensuring the place of science and engineering on the nation's list of priorities.

Augustine served for 16 years as a member of the President's Council of Advisors on Science and Technology and is currently on the advisory councils of the U.S. Department of Homeland Security and the Department of Energy. He was among several individuals who testified to Congress regarding the National Academies' report *Rising Above the Gathering Storm: Energizing and Employing America for a Brighter Economic Future*. Augustine chaired the panel that conducted the study, which was requested by Congress. The report recommends ways to strengthen research and education in science and technology.

Among Augustine's many honors are the National Medal of Technology and the U.S. Department of Defense's highest civilian award, the Distinguished Service Medal, given to him five times. Most recently, he was awarded the 2005 American Association for the Advancement of Science (AAAS) Philip Hauge Abelson Prize and the 2006 Public Welfare Medal from the National Academy of Sciences.

Augustine served as chairman and principal officer of the American Red Cross for 9 years and as chairman of the NAE, the Association of the United States Army, the Aerospace Industries Association, and the Defense Science Board. He is a former president of the American Institute of Aeronautics and Astronautics and the Boy Scouts of America. He is a former member of the board of directors of ConocoPhillips, Black & Decker, Procter & Gamble, and Lockheed Martin and of the board of trustees of Colonial Williamsburg, a trustee emeritus of Johns Hopkins University, and a former member of the Board of Trustees of Princeton and MIT. He holds 28 honorary degrees. He is the author or co-author of *Augustine's Travels*, *The Defense Revolution*, *Augustine's Laws*, and *Shakespeare in Charge*.

Born in Colorado in 1935, Augustine attended East Denver High School and graduated magna cum laude from Princeton University, where he earned bachelor's and master's degrees in aeronautical engineering.

C.D. (Dan) Mote, Jr. (NAE), *Co-chair,* is Regents Professor and Glenn L. Martin Institute Professor of Engineering at the University of Maryland, where he was president of the university from 1998 to 2010. Under his leadership, academic programs flourished, leading the university to a 36th in the world ranking by the Academic

Ranking of World Universities. Mote is a leader in the national dialogue on higher education, and his analyses of shifting funding models have been featured in local and national media. He has testified on major educational issues before Congress, representing the University of Maryland and higher education associations on the problem of visa barriers for international students and scholars, global competitiveness, and deemed export control issues. He has served or currently serves on National Research Council (NRC) committees that work to identify challenges to U.S. leadership in key areas of science and technology. He chaired the 2010 NRC study *S&T Strategies of Six Countries: Implications for the United States*; served as vice chair of the U.S. Department of Defense Basic Research Committee; is a member and officer of the NAE; co-chairs the Government-University-Industry Research Roundtable; and serves on the governing board of the NRC.

In 2004-2005, he served as president of the Atlantic Coast Conference. In its last ranking in 2002, "Washington Business Forward" magazine counted him among the 20 most influential leaders in the region. Before assuming the presidency at Maryland, Mote served on the faculty of the University of California, Berkeley, for 31 years. From 1991 to 1998, he was vice chancellor at Berkeley, held an endowed chair in mechanical systems, and was president of the UC Berkeley Foundation. He led a comprehensive capital campaign for Berkeley that ultimately raised $1.4 billion. He earlier served as chair of UC Berkeley's Department of Mechanical Engineering and led the department to its number one ranking in the National Research Council review of graduate program effectiveness.

Mote is internationally recognized for his research on the dynamics of gyroscopic systems and the biomechanics of snow skiing and has produced more than 300 publications. He holds patents in the United States, Norway, Finland, and Sweden and has mentored 58 PhD students. Mote has received numerous awards and honors, including the Humboldt Prize, awarded by the Federal Republic of Germany. He is a recipient of the Berkeley Citation from the University of California and was named Distinguished Engineering Alumnus. He has received three honorary doctorates, is a member of the U.S. National Academy of Engineering, a fellow of the American Academy of Arts and Sciences, an honorary member of the American Society of Mechanical Engineers (ASME) International, and is fellow of the International Academy of Wood Science, the Acoustical Society of America, and the American Association for the Advancement of Science. In the spring of 2005, he was named a recipient of the J.P. Den Hartog award by the ASME International to honor his lifelong contribution to the teaching and/or practice of vibration and sound. He received the 2005 Founders Award from the National Academy of Engineering in recognition of his comprehensive body of work on the dynamics of moving flexible structures and for leadership in academia. He received BS, MS, and PhD degrees in mechanical engineering from the University of California, Berkeley.

Burt S. Barnow is the Amsterdam Professor of Public Service and Economics at the Trachtenberg School of Public Policy and Public Administration at George Washington University. He has over 30 years of experience as an economist in the fields of workforce investment, program evaluation, performance analysis, labor economics, welfare, poverty, child support, and fatherhood programs. Before coming to George Washington University, Barnow was associate director for research at Johns Hopkins University's Institute for Policy Studies, where he worked for 18 years. Prior to that, he had worked for 8 years at the Lewin Group and nearly 9 years at the U.S. Department of Labor, including 4 years as director of the Office of Research and Evaluation in the Employment and Training Administration. Even earlier, Barnow was an assistant professor of economics at the University of Pittsburgh. He has extensive experience conducting research on implementation of large government programs and is currently co-project-director for a study for the Employment and Training Administration (ETA) to analyze states' experiences in implementing workforce investment and unemployment insurance provisions of the Recovery Act. Barnow also co-directed studies for ETA on the implementation of the Workforce Investment Act (WIA) and the 1992 amendments to the Job Training Partnership Act (JTPA). His current and recent research includes an evaluation of the Center for Working Families programs for the Annie E. Casey Foundation, a project to develop and evaluate demonstrations that test innovative strategies to promote self-sufficiency for low-income families for the U.S. Department of Health and Human Services, a study for ETA to evaluate the impact of selected projects in the High Growth Job Training Initiative using nonexperimental methods, an assessment of occupational skill shortages for the Alfred P. Sloan Foundation, an evaluation of the priority of services for the veterans' mandate for Department of Labor programs for ETA, a project to develop cost-performance standards for ETA, an evaluation of the determinants of the welfare caseload in Colorado for the State of Colorado, and an evaluation of a

Department of Labor demonstration project to help youth in foster care make the transition into the labor market for Casey Family Programs.

Barnow served as vice chairman of the National Research Council's Committee on the Information Technology Work Force and was a member of the Board on Higher Education and Workforce for 6 years. He is currently serving on the NRC Committee on the External Evaluation of the National Institute on Disability and Rehabilitation Research and the Committee on the U.S. Mining and Energy Workforce, and he has served on five other NRC committees. He currently serves on the Baltimore Workforce Investment Board's System Effectiveness Committee, and he chaired the Performance Committee of the Maryland Governor's Workforce Investment Board for 4 years. Barnow chairs the National Association of Schools of Public Affairs Research Committee, and he serves on the editorial boards of two journals. He has a BS degree in economics from the Massachusetts Institute of Technology and MS and PhD degrees in economics from the University of Wisconsin at Madison.

James S.B. Chew is L-3 Communications Holdings director of advanced technologies and concepts for the Precision Engagement Sector. Chew is responsible for leading the development and transition of disruptive precision engagement technologies to the DOD and commercial markets. He is also chairman of the Science and Engineering Technology Division of the National Defense Industrial Association. Prior to joining L-3, Chew served as a propulsion engineer for Boeing Aerospace Company; senior engineer for SPARTA; program manager for the Air Force Rocket Propulsion Lab; director of rocket propulsion technology plans and programs for the Air Force Phillips Laboratory; assistant staff specialist for weapons technology for the Office of the Secretary of Defense; and deputy director of air and surface weapons technology for the Office of Naval Research. Chew also served as Exide's vice president for the Military and Specialty Global Business Unit; product marketing consultant for the Dodge Division of Chrysler Corporation; QWIPTECH's chief operating officer; General Motors' American Tuner program manager; T/J Technologies chief operating officer; vice president, science and technology, ATK; and SAIC's vice president for the Space Systems Development Division. Chew earned a lifetime California State Community College teaching credential in engineering. He serves on the board of ABAKAN, Inc. Chew is a graduate of the Stanford Executive Engineering Program and the Defense Systems Management College Advanced Program Management Program. He is a DOD Level 3 certified acquisition professional and a DOD Level 3 System, Planning, Development, Research, and Engineering professional. He was recognized as the 2009 College of Engineering Distinguished Alumnus by his undergraduate alma mater. He earned his BS degree in mechanical engineering from the California State Polytechnic University at Pomona and an MS degree in systems management from the University of Southern California.

Lawrence J. Delaney retired as the executive vice president of operations and president of the Advanced Systems Development Sector of Titan Corporation. Previously, he held distinguished positions with Arete Associates, Inc.; Delaney Group, Inc.; BDM Europe; and the Environmental and Management Systems Group at IABG. Delaney was also the acting secretary of the Air Force and served as the assistant secretary of the Air Force for acquisition, as well as the Air Force's service acquisition executive, responsible for all Air Force research, development, and acquisition activities. He provided direction, guidance, and supervision for all matters pertaining to the formulation, review, approval, and execution of acquisition plans, policies, and programs. Delaney has more than 41 years of international experience in high-technology program acquisition, management, and engineering, focusing on space and missile systems, information systems, propulsion systems, and environmental technology. He served as a member of the National Research Council's Board on Army Science and Technology (vice chair), Air Force Studies Board (chair), and the Army Science Board (vice chair). Delaney received his bachelor's and master's degrees in chemical engineering from Clarkson University and his PhD in chemical engineering from the University of Pennsylvania.

Mary L. Good (NAE) is dean emeritus and special advisor to the chancellor at the University of Arkansas at Little Rock. She is managing member for the Fund for Arkansas' Future, LLC (an investment fund for start-up and early-stage companies), past president of the AAAS, past president of the American Chemical Society, and an elected member of the National Academy of Engineering. She currently serves on the boards of St. Vincent Health

System and Delta Bank and Trust. Previously she served a 4-year term as the under secretary for technology for the Technology Administration in the Department of Commerce, appointed by the President and confirmed by the Senate. In addition, she chaired the National Science and Technology Council's Committee on Technological Innovation (NSTC/CTI) and served on the NSTC Committee on National Security. Previously she served as the senior vice president for technology for Allied Signal and as the Boyd Professor of Chemistry and Materials Science at Louisiana State University. She was appointed to the National Science Board by President Carter in 1980 and by President Reagan in 1986. She was the chair of that board from 1988 until 1991, when she was appointmented by President Bush to be a member of the President's Council of Advisors on Science and Technology. She has received many awards, including the National Science Foundation's Distinguished Public Service Award, the American Institute of Chemists' Gold Medal, the Priestly Medal from the American Chemical Society, and the Vannevar Bush Award from the National Science Board, among others. Good received her bachelor's degree in chemistry from the University of Central Arkansas and her MS and PhD in inorganic chemistry from the University of Arkansas at Fayetteville.

Daniel E. Hastings is the Cecil and Ida Green Education Professor of Aeronautics and Astronautics and Engineering Systems at the Massachusetts Institute of Technology as well as the dean for undergraduate education. Hastings has taught courses and seminars in plasma physics, rocket propulsion, advanced space power and propulsion systems, aerospace policy, technology and policy, and space systems engineering. Hastings served as chief scientist to the U.S. Air Force from 1997 to 1999. In that role, he acted as chief scientific adviser to the chief of staff and the secretary and provided assessments on a wide range of scientific and technical issues affecting the Air Force mission. He led several influential studies advising the Air Force investment in space, global energy projection, and options for a science and technology workforce for the twenty-first century. His recent research has concentrated on issues of space systems and space policy and also on issues related to spacecraft environmental interactions, space propulsion, and space systems engineering. He has published many papers and a book in the field of spacecraft-environment interactions and several papers in space propulsion and space systems. He has also led several national studies on government investment in space technology. Hastings is a fellow of the American Institute of Aeronautics and Astronautics, a fellow of the International Council on Systems Engineering, and a full academician of the International Academy of Astronautics. He served as a member of the National Science Board and the Applied Physics Lab Science and Technology Advisory Panel, as well as the chair of the Air Force Scientific Advisory Board. He is a member of the MIT Lincoln Laboratory Advisory Committee, a member of the corporation of Draper Laboratory, and a member of the board of trustees for the Aerospace Corporation. He has served on several national committees on issues in the national security space. As dean for undergraduate education, Hastings has broad responsibility for policy and direction in undergraduate education at MIT. He also oversees several administrative offices at MIT, including the Office of Undergraduate Advising and Academic Programming, Admissions Office, Global Education and Career Development Center, Office of Experiential Learning, Office of Educational Innovation and Technology, Office of Faculty Support, Office of Minority Education, Registrar's Office, Student Financial Services, the Teaching and Learning Laboratory, and the ROTC programs. Hastings earned a BA in mathematics from Oxford University and a PhD and an SM in aeronautics and astronautics from MIT.

Robert J. Hermann (NAE) is a private consultant. Previously he served as a senior partner at Global Technology Partners, LLC. He retired as senior vice president for science and technology of the United Technologies Corporation in 1998. He is a former director of the Defense Department's National Reconnaissance Office and a former senior official at the National Security Agency. Hermann served as a member of the President's Foreign Intelligence Advisory Board (1993-1995) during the William Jefferson Clinton Administration. In 1998, he retired from United Technologies Corporation, where he held the position of senior vice president for science and technology. In that role, he was responsible for ensuring the development of technical resources and the full exploitation of science and technology by the corporation. He was also responsible for the United Technologies Research Center. Hermann joined the company in 1982 as vice president of systems technology in the electronics sector and later served in a series of assignments in the defense and space systems groups prior to being named vice president of science and technology. Hermann concluded his tenure as immediate past chairman of the American National Standards Institute (ANSI) board of directors at the end of 2002 following a 2-year term; he had served as chairman of the ANSI board of directors during 1999 and 2000 and as a member of the ANSI board since 1993. Prior to joining UTC, Hermann served for 20 years with the National Security Agency, with assignments in research and development, in operations, and at NATO. In 1977, he was appointed principal deputy assistant secretary of defense for communications, command, control and intelligence. In 1979, he was named assistant secretary of the Air Force for research, development, and logistics and in parallel was director of the National Reconnaissance Office. He received BS, MS, and PhD degrees in electrical engineering from Iowa State University.

J.C. Herz is the CEO of Batchtags, Inc. She is a technologist with a background in biological systems and computer game design. Her specialty is massively multiplayer systems that leverage social network effects, whether on the Web, mobile devices, or more exotic high-end or grubby low-end hardware. She currently serves as a White House special consultant to the Office of the Secretary of Defense (Networks and Information Integration). Defense projects range from aerospace systems to a computer-game-derived interface for next-generation unmanned air systems. Herz is one of the three co-authors of OSD's Open Technology Development roadmap. She serves on the Federal Advisory Committee for the National Science Foundation's education directorate. In that capacity, she is helping NSF harness emerging technologies to drive U.S. competitiveness in math and science. Herz was a member of the NRC's committee on IT and Creative Practice and is currently a fellow of Columbia University's American Assembly, where she is on the leadership team of the assembly's Next Generation Project. In 2002, she was designated a Global Leader for Tomorrow by the World Economic Forum. She is a member of the Global Business Network, a founding member of the Institute of Electrical and Electronics Engineers (IEEE) Task Force on Game Technologies, and a term member of the Council on Foreign Relations. She is on the advisory board of Carnegie Mellon's ETC Press and is the author of two books, *Surfing on the Internet* (Little, Brown, 1994), an ethnography of cyberspace before the Web, and *Joystick Nation: How Videogames Ate Our Quarters, Won Our Hearts, and Rewired Our Minds* (Little, Brown, 1997), a history of videogames that traces the cultural and technological evolution of the first medium that was born digital, and how it shaped the minds of a generation weaned on Nintendo. Her books have been translated into seven languages. As a *New York Times* columnist, Herz published 100 essays between 1998 and 2000 on the grammar and syntax of game design. She has also contributed to Esther Dyson's *Release 1.0, Rolling Stone, Wired, GQ*, and the *Calgary Philatelist*. Herz graduated magna cum laude from Harvard with a B.A. in biology and environmental studies.

Ray O. Johnson, a global executive focused on innovation and diversity, is the senior vice president and chief technology officer of the Lockheed Martin Corporation. As an officer of the corporation and a member of the executive leadership team, Johnson guides the corporation's technology vision and provides corporate leadership in the strategic areas of technology and engineering, which have more than 65,000 people working on more than 4,000 programs that provide some of the nation's most vital security systems. He has a proven track record in managing large profit and loss organizations, strategic planning, program development, program management, and venture capital funding.

Johnson currently serves as a member of the boards of directors of Sandia Corporation, the National Math and Science Initiative, and the Hispanic College Fund. He is a member of the governing board of the Indo-U.S. Science and Technology Forum and a sponsor of the DST-Lockheed Martin India Innovation Growth Program. Johnson is on the U.S. National Institute of Standards and Technology (NIST) Technology Innovation Program Advisory Board. He is on the board of directors of the Northern Virginia Technology Council and the Virginia Center for Innovative Technology. He is also a member of the Virginia Innovation and Entrepreneurship Investment Authority and the Maryland Federal Facilities Advisory Board. He is a member of the board of visitors for the A. James Clark School of Engineering at the University of Maryland, on the dean's advisory council for the College of Engineering at Carnegie Mellon University, and a member of the board of affiliates of the Rice University Professional Science Master's Program. He is also the chairman of the USA Science and Engineering Festival's advisory board, which held its inaugural event in October 2010 on the National Mall in Washington, D.C., and was attended by more than 1 million people. He is a full academician of the International Academy of

Astronautics (IAA) and a fellow of the International Society for Optical Engineering (SPIE), the American Institute of Aeronautics and Astronautics (AIAA), and the Institute of Electrical and Electronics Engineers. He is a member of Eta Kappa Nu, Tau Beta Pi, and Phi Kappa Phi. Johnson also chairs the Technology Leadership and Strategy Initiative of the U.S. Council on Competitiveness. He holds BS (Oklahoma State University), MS, and PhD (Air Force Institute of Technology) degrees in electrical engineering.

Anita K. Jones (NAE) is a university professor emerita at the University of Virginia and a professor of computer science in the School of Engineering and Applied Science, previously having served as chair of the Department of Computer Science. Jones was sworn in as the director of defense research and engineering for the U.S. Department of Defense in 1993. In that position she was responsible for the management of the DOD science and technology program. This included responsibility for the Defense Advanced Research Projects Agency and oversight of the DOD laboratories, as well as being the principal advisor to the secretary of defense for defense-related scientific and technical matters. Jones is past vice chair of the National Science Board, which advises the President on science, engineering, and education and oversees the National Science Foundation. She is a senior fellow of the Defense Science Board, a member of the Charles Stark Draper Laboratory Corporation, and a past member of the MIT Corporation Executive Committee. She has co-chaired the Commonwealth of Virginia Research and Technology Advisory Commission and has served on other government advisory boards and scientific panels for NASA, the National Academies, the Department of Energy, and the National Science Foundation. She is a member of the National Academy of Engineering and the American Philosophical Society, and a fellow of the Computing Research Association (CRA). She is a recipient of the CRA's Service Award, the Air Force Meritorious Civilian Service Award, the Department of Defense Award for Distinguished Public Service, and the IEEE Founders Award. The U.S. Navy named a seamount in the North Pacific Ocean for her. She is currently a member of the board of directors of Science Applications International Corporation and of ATS Corporation and is a trustee of In-Q-Tel. Other private sector experience includes serving as a trustee of the MITRE Corporation. Duke University, Carnegie Mellon University, and the University of Southern California have awarded her honorary doctorate degrees. She is a founder and council member of the Computing Community Consortium. She has published more than 50 technical articles and two books in the area of computer software and systems, cybersecurity, and science and technology policy. In the fall of 2010, the National Academy of Engineering gave her the Arthur M. Bueche Award for contributions to science and technology policy advancement. Jones holds an AB from Rice University in mathematics, an MA from the University of Texas, Austin, in literature, and a PhD in computer science from Carnegie Mellon University.

Sharon Levin is professor emeritus and research professor of economics at the University of Missouri-St. Louis. Levin has been studying issues concerning the science and engineering workforce for more than 25 years. She has co-authored the book *Striking the Mother Lode in Science: The Importance of Age, Place, and Time* (1992), and her work related to the science and engineering workforce has been published in such prominent journals as the *American Economic Review, The Review of Economics and Statistics, Growth and Change, Science, Social Studies of Science, and Management Science*. Her research on the careers of scientists and engineers has also been the focus of articles in *Economist, Science, Scientist,* and various newspapers and magazines in the United States and abroad. In 1993 she was awarded the Chancellor's Award for Excellence in Research and Creativity by the University of Missouri-St. Louis. Levin's research currently focuses on the effects that the diffusion of information technology has had on the publishing productivity of academic scientists. Her research has been supported by the Alfred P. Sloan Foundation, the Exxon Educational Foundation, the Andrew W. Mellon Foundation, the National Science Foundation, the National Bureau of Economic Research, and the University of Missouri. Levin graduated from the Bronx High School of Science and the City College of New York (Phi Beta Kappa, magna cum laude) with a BA in economics, and she earned both an MA and a PhD in economics from the University of Michigan.

Frances S. Ligler (NAE) is the Navy's senior scientist for biosensors and biomaterials and a past chair of the Bioengineering Section of the NAE. Currently working in the fields of biosensors and microfluidics, she has also performed research in biochemistry, immunology, and proteomics. She has over 350 full-length publications and patents, which have led to 11 commercial biosensor products and have been cited over 7,500 times. She is a winner of the Navy Superior Civilian Service Medal, the National Drug Control Policy Technology Transfer Award, the Chemical Society Hillebrand Award, the Navy Merit Award, the Naval Research Laboratory (NRL) Technology Transfer Award, three NRL Edison Awards for Patent of the Year, the Furman University Bell Tower and Distinguished Alumni of the 20th Century Awards, and the national Women in Science and Engineering (WISE) Outstanding Achievement in Science award. She serves as an associate editor of *Analytical Chemistry* and is on the editorial/advisory boards for *Biosensors & Bioelectronics*, *Analytical Bioanalytical Chemistry*, *Sensors*, *Open Optics*, and *Applied Biochemistry and Biotechnology*. Elected an SPIE fellow in 2000 and a fellow of the American Institute for Medical and Biological Engineering in 2011, she also serves on the organizing committee for the World Biosensors Congress and the permanent steering committee for Europt(r)odes, a European conference on optical sensors. In 2003, she was awarded the Homeland Security Award (Biological, Radiological, Nuclear Field) by the Christopher Columbus Foundation and the Presidential Rank of Distinguished Senior Professional by President Bush. She earned a BS from Furman University and a DPhil and a DSc from Oxford University.

Aaron Lindenberg is an assistant professor of materials science and engineering at Stanford University, where his current research is focused on the dynamics of phase transitions, ultrafast properties of nanoscale materials, photoelectrochemical charge transfer dynamics, and terahertz nonlinear spectroscopy. Prior to his current position he served as a staff scientist at the Stanford Synchrotron Radiation Laboratory. Previous to that assignment he was a postdoctoral faculty fellow at the University of California, Berkeley. In 2010 he was named a DARPA young faculty awardee for functional materials—all-optical control of nanoelectronic devices. From 2007 to 2009 he was a Stanford Terman Fellow. He won the Alfred Moritz Michaelis Prize in physics from Columbia University as well as being the I.I. Rabi Scholar while at Columbia. He received his BA from Columbia University and his PhD at the University California, Berkeley.

Paul D. Nielsen (NAE) is director and chief executive officer of the Software Engineering Institute (SEI), a federally funded research and development center operated by Carnegie Mellon University. The SEI advances software engineering and cybersecurity principles and practices through focused research and development, which is transitioned to the broad software engineering community. Prior to his arrival as SEI director, Nielsen served in the U.S. Air Force, retiring as a major general after 32 years of distinguished service. As commander of the Air Force Research Laboratory at Wright-Patterson Air Force Base in Ohio for more than 4 years, he managed the Air Force's science and technology budget of more than $3 billion annually. He also served as the Air Force's technology executive officer, determining the investment strategy for the full spectrum of Air Force science and technology activities. Nielsen is a member of the National Academy of Engineering, a fellow of the AIAA, and a fellow of the IEEE. He served as the AIAA president in 2007 and 2008 and was on the AIAA board from 2006 to 2009. Nielsen has served on several technical advisory committees and boards, including the Air Force Scientific Advisory Board. He is a member of the board of directors for the Hertz Foundation, a non-profit that awards graduate school fellowships in the applied sciences. Nielson received a BS in physics and mathematics from the U.S. Air Force Academy; an MS in applied science from the University of California, Davis; an MBA from the University of New Mexico; and a PhD in plasma physics from the University of California, Davis.

Daniel T. Oliver USN (Vice Admiral, ret.) is the president of the Naval Postgraduate School. Commissioned in 1966 through the Naval Reserve Officer Training Corps program at the University of Virginia, he became a naval aviator and piloted the Navy's P-3 maritime patrol aircraft, specializing in detecting and tracking submarines. He completed eight operational deployments around the world during the Cold War, commanding Patrol Squadron Sixteen and Patrol Wing Two. As a flag officer, he served as commander, Fleet Air Forces Mediterranean, and commanded coalition air operations in support of the United Nations' embargo of the former Republic of Yugoslavia. Oliver served on the personal staffs of two chiefs of naval operations. In his first flag assignment as director, Total Forces Training and Education Division, he supervised mobilization of naval reservists called to active duty during Operation Desert Storm. He later served sequentially as director of the OPNAV Assessment Division, the Fleet Liaison Division, and the Programming Division. In these capacities, he was instrumental in shaping a balanced investment program for all Navy resources during the post-Cold-War drawdown. In September 1996, Oliver became the chief of naval personnel and deputy chief of naval operations for manpower and personnel. He was the primary advocate for sailors, both officers and enlisted men and women, from recruitment through retirement. In this position, he formulated and instituted personnel policies that guided the Navy through a critical transition from a post-Cold-War drawdown to a steady-state force. After retiring from active duty in 2000, he was active in the private sector as a senior executive and board member of a number of companies and civic organizations, mostly involved with government contracting in the information technology sector. Oliver is a graduate of the Harvard Business School Advanced Management Program and was a White House Fellow. He holds a bachelor's and a master's degree from the University of Virginia, where he also served as an associate professor of naval science.

C. Kumar N. Patel (NAS/NAE) is the founder, president, and CEO of Pranalytica, Inc., a Santa Monica-based company that is the leader in quantum cascade laser technology for defense and homeland security applications. He is also a professor of physics and astronomy, electrical engineering, and chemistry at the University of California, Los Angeles (UCLA). He served as vice chancellor for research at UCLA from 1993 to 1999. Prior to joining UCLA in March 1993, he was the executive director of the Research, Materials Science, Engineering and Academic Affairs Division at AT&T Bell Laboratories, Murray Hill, New Jersey. He joined Bell Laboratories in 1961, where he began his career by carrying out research in gas lasers. He is the inventor of the carbon dioxide laser and many other molecular gas lasers that ushered in the era of high-power sources of coherent optical radiation. In 1996, Patel was awarded the National Medal of Science by the President. His other awards include the Ballantine Medal of the Franklin Institute, the Zworykin Award of the National Academy of Engineering, the Lamme Medal of the Institute of Electrical and Electronics Engineers, the Texas Instruments Foundation Founders' Prize, the Charles Hard Townes Award of the Optical Society of America, the Arthur H. Schawlow Award of the Laser Institute of America, the George E. Pake Prize of the American Physical Society, the Medal of Honor of the IEEE, the Frederic Ives Medal of the Optical Society of America, and the William T. Ennor Manufacturing Technology Award of the ASME.

He is a member of the board of directors of the Newport Corporation. He served on the board of trustees of the Aerospace Corporation from 1979 to 1989. He is a member of the National Academy of Sciences and the National Academy of Engineering. He has served on several NRC committees, including the Committee on the Navy's Needs in Space for Providing Future Capabilities, the Air Force Studies Board, the Panel on Sensors and Electron Devices, and the congressionally mandated NRC Committee on an Assessment of Concepts and Systems for U.S. Boost Phase Missile Defense in Comparison with Other Alternatives. He is currently co-chairing the NRC Committee to Review the Quality of Science and Engineering Research at the National Security Labs. In 1988, he was awarded an honorary DSc degree from the New Jersey Institute of Technology. Patel holds a BE degree in telecommunications from the College of Engineering in Poona, India, and received an MS and a PhD degree in electrical engineering from Stanford University.

Leif E. Peterson is managing partner for Advanced HR Concepts and Solutions. Before retiring in 2007, Peterson was a member of the Senior Executive Service and the director of Manpower, Personnel, and Services for the Air Force Materiel Command (AFMC) at Wright-Patterson Air Force Base in Ohio. He provided executive management of the command's nearly 80,000 military and civilian professionals throughout the United States and overseas in research facilities, test sites, and universities and at product development, logistics, and specialized centers. The function of the Directorate of Manpower, Personnel, and Services was to shape the AFMC workforce to deliver war-winning expeditionary capabilities and provide oversight, direction, and control for all personnel activities within AFMC. Peterson entered federal service in 1971 as a labor relations specialist at the U.S. Air Force Headquarters. He held numerous positions as a civilian personnel officer, serving two tours at Eglin Air Force Base in Florida and 6 years overseas. In 1983, Peterson became deputy director of civilian personnel for Air Force Systems Command at Andrews Air Force Base in Maryland. He later returned to U.S. Air Force Headquarters as chief of staffing, development, and equal employment opportunity. For 8 years he was director of civilian personnel at the Tactical Air Command and Air Combat Command at Langley Air Force Base in Virginia. He was then assigned as director of civilian personnel and programs at AFMC. He was appointed to the Senior Executive Service in 2004, assuming his previous position as deputy director of personnel. He received a BS in labor and industrial relations from Michigan State University and an MS in labor and industrial relations from Loyola University.

Stephen M. Robinson (NAE) is professor emeritus of industrial and systems engineering and of computer sciences at the University of Wisconsin-Madison, where he served on the faculty from 1972 to 2007. Robinson also holds the rank of colonel (retired) in the U.S. Army. His research specialty is variational analysis and mathematical programming: methods for making the best use of limited resources, applied to logistics, transportation, manufacturing, and many other areas. He is author, coauthor, or editor of seven books and more than 100 scientific research papers and has directed numerous funded research projects at the university. His research accomplishments have been recognized by the award of an honorary doctorate from the University of Zürich, Switzerland, the George B. Dantzig Prize of the Mathematical Programming Society and the Society for Industrial and Applied Mathematics (SIAM), and the John K. Walker Jr. Award of the Military Operations Research Society. He is a member of the National Academy of Engineering, a national associate of the National Research Council, a fellow of the Institute for Operations Research and the Management Sciences, and a fellow of the SIAM. He received a BA in mathematics from the University of Wisconsin, an MS in mathematics from New York University, and a PhD in computer sciences from the University of Wisconsin.

Michael S. Teitelbaum is the Wertheim Fellow at Harvard Law School. He is also a senior advisor to the Alfred P. Sloan Foundation. By specialty he is a demographer, with research interests in the causes and consequences of very low fertility rates; the drivers and implications of international migration; and science and engineering labor markets. He has written and edited 10 books and many articles on these subjects. Previously he served as vice president of the Sloan Foundation; faculty member at Oxford and Princeton universities; director of the U.S. Congressional Select Committee on Population; vice chair and acting chair of the U.S. Commission on International Migration; member of the U.S. Commission on International Migration and Cooperative Economic Development; and chair of the Section on Social, Economic and Political Sciences of the AAAS, of which he was later elected a fellow. Teitelbaum was educated at Reed College and at Oxford University, where he was a Rhodes Scholar. He earned a DPhil from Oxford University.

Ronald Williams is a vice president of the College Board. Among several leadership roles, Williams is responsible for strengthening the relationship between the College Board and community colleges throughout the United States. He also provides leadership to a cluster of initiatives dealing with students' access to, and persistence in, college. Williams joined the College Board in 2007 from Prince George's Community College in Largo, Maryland, where he had served as president since 1999, capping an extensive career with community colleges. Williams is a member of the board of the American Association of Colleges and Universities, the American Association of Community Colleges, and the American Council on Education's Center for Policy Analysis Advisory Committee. A writer, Williams has published two novels, *Four Saints and an Angel* and *A Death in Panama*. Williams received a bachelor's degree in history and English, a master's degree in English, and a doctorate in literature from Lehigh University.

Appendix B

Meetings and Speakers

MEETING 1
JUNE 7-8, 2011
HOTEL PALOMAR
WASHINGTON, D.C.

STEM Workforce Development for the Department of Defense
Zachary Lemnios, Assistant Secretary of Defense for Research and Engineering
Office of the Assistant Secretary of Defense for Research and Engineering

STEM Education, Skills, Capabilities and Capacity Needs for the Department of Defense
Laura Adolfie, Director, STEM Development Office
Office of the Assistant Secretary of Defense for Research and Engineering

Air Force STEM Study
Daniel Talmage, Program Officer
Air Force Studies Board
National Research Council

Assessment of Civilian Science and Engineering Workforce in DOD Laboratories
Jocelyn M. Seng, Research Staff Member, Science and Technology Division
Institute for Defense Analyses

Exploration of the Department of Defense's Civilian Acquisition Workforce
Susan M. Gates, Director
Kauffman-RAND Institute for Entrepreneurship Public Policy

WORKSHOP ON SCIENCE, TECHNOLOGY, ENGINEERING AND MATHEMATICS (STEM) WORKFORCE NEEDS FOR THE U.S. DEPARTMENT OF DEFENSE AND THE U.S. DEFENSE INDUSTRIAL BASE
AUGUST 1-2, 2011
THE WATERVIEW CONFERENCE CENTER
ARLINGTON, VIRGINIA

STEM Workforce Needs for U.S. DOD and Defense Industry Base
Charles M. Vest, President
National Academy of Engineering

Purpose and Plan
Norman R. Augustine and C.D. (Dan) Mote, Committee Co-chairs
Committee on Science, Technology, Engineering, and Mathematics (STEM) Workforce Needs for the U.S. Department of Defense and the U.S. Defense Industrial Base

Building the Future Technical Workforce for the U.S. Department of Defense and the U.S. Defense Industrial Base
Zachary J. Lemnios, Assistant Secretary of Defense Research & Engineering
Office of the Assistant Secretary of Defense Research & Engineering

Panel I: Emerging Science and Technology in the Next 15 years

Introductory Talks

Emerging Science and Technology in the Life Sciences
Donald Burke, Dean, Graduate School of Public Health
University of Pittsburgh

S&T That Will Impact DOD Over the Next 15 Years
Anthony Tether, President
The Sequoia Group
Distinguished Fellow, Council on Competitiveness

Panel I Discussion
Moderator: Frances Ligler
U.S. Naval Research Laboratory

Thomas Russell, Director
Air Force Office of Scientific Research

Science, Technology, Engineering, and Mathematics Workforce Needs for the U.S. Department of Defense and the U.S. Defense Industrial Base
Lyle Schwartz
American Society for Materials, Materials Educational Foundation

Emerging Science and Technology in Next 15 Years
John Sommerer, Space Department Head
Johns Hopkins University Applied Physics Laboratory

Leonard Buckley, Director, Science and Technology Division
Institute for Defense Analyses

Panel II: Estimating Science, Technology, Engineering, and Mathematics (STEM) Workforce Needs Under Future Scenarios

Introductory Talk

STEM Workforce Needs of the U.S. Department of Defense: Background Data
Rolf Lehming
Director, Science and Engineering Indicators
National Center for Science and Engineering Statistics
National Science Foundation

Panel II Discussion
Moderator: Anita Jones
University of Virginia

Selected Studies
Leif Peterson, Managing Partner
Advanced HR Concepts & Solutions

Data Sources on the STEM Workforce
Dixie Sommers, Assistant Commissioner of Labor Statistics
U.S. Department of Labor

National Academies Workshop "Estimating STEM Workforce Needs Under Future Scenarios"
John Fischer, Director, Laboratories Office
Office of the Assistant Secretary of Defense Research and Engineering

NDIA STEM Workforce Division
Edward Swallow, Vice President, Business Development
Northrop Grumman Corporation
Chairman, STEM Workforce Division
National Defense Industrial Association

Panel III: Limitations to Meeting Workforce Needs of Department of Defense (DOD) and the Industrial Base

Introductory Talk

The New STEM Labor Market Segmentation: Implications for Meeting Workforce Needs of DoD and the Industrial Base
Harold Salzman, Professor of Public Policy
Rutgers University

Panel III Discussion
Moderator: Sharon Levin
University of Missouri-St. Louis

What Are Labor Shortages and How Do They Arise?
Burt Barnow, Amsterdam Professor of Public Service and Economics, Trachtenberg
School of Public Policy and Public Administration
George Washington University

Employment in STEM Occupations
Dixie Sommers, Assistant Commissioner of Labor Statistics
U.S. Department of Labor

Stay Rates of Foreign Doctorate Recipients
Mike Finn, Economist
Oak Ridge Institute for Science Education

Rick Stephens, Senior Vice President Human Resources and Administration
The Boeing Company

Panel IV: Institutional Capacity in Education and the DOD Investments Needed to Ensure a Sufficient Workforce

Introductory Talk

Creating a More STEM Capable DOD Workforce
Carl Wieman; Associate Director for Science
Office of Science and Technology Policy

Panel IV Discussion
Moderator: Daniel Oliver
Naval Postgraduate School

Katrina McFarland, President
Defense Acquisition University

STEM Workforce Needs of DOD and the U.S. Defense Industrial Base
Wes Harris, Charles Stark Draper Professor of Aeronautics and Astronautics
Associate Provost
Massachusetts Institute of Technology

Do We Have Capacity?
Paul Gaffney, President
Monmouth University

The Undereducated American
S. James Gates, Jr.
John S. Toll Professor of Physics, and Director of Center for String and Particle Theory University of Maryland, College Park

Panel V: Ensuring an Adequate Workforce Capability in an Uncertain Future

Introductory Talk

DOD STEM Planning in Uncertain Times
Ruth David, President
ANSER

Panel V Discussion
Moderator: Robert Hermann
National Academy of Engineering

David Chu, President
Institute for Defense Analyses

Jennifer Byrne, Vice President, Corporate Engineering and Technology
Lockheed Martin

Vallen Emery, Outreach Program Manager
U.S. Army Research Laboratory

Katherine McGrady, Executive Vice President and Chief Operating Officer
CNA

MEETING 2
AUGUST 3, 2011
KECK CENTER OF THE NATIONAL ACADEMIES
WASHINGTON, D.C.

Closed Meeting

MEETING 3
SEPTEMBER 18-19, 2011
KECK CENTER OF THE NATIONAL ACADEMIES
WASHINGTON, D.C.

Panel Discussion of Service-Specific S&T Needs

Science, Technology, Engineering and Mathematics (STEM) in the Air Force
Mark Maybury, Chief Scientist
U.S. Air Force

DoN STEM Efforts
Larry Schuette, Director of Innovation
U.S. Office of Naval Research

STEM: It's Not Just Science, Technology, Engineering and Mathematics
Marilyn Freeman, Deputy Assistant Secretary
U.S. Army

Scott Fish, Chief Scientist
U.S. Army

Presentations

DOD STEM Workforce: Government, Industry and Academia
Jacques Gansler, Professor and Roger C. Lipitz Chair
Director, Center for Public Policy and Private Enterprise, School of Public Policy
University of Maryland

The National Security Imperative for Global S&T Engagement
Gerald Epstein, Center for Science, Technology and Security Policy
American Association for the Advancement of Science

AFMC Commander's Perspective on STEM
General Donald J. Hoffman
Air Force Materiel Command

MEETING 4
NOVEMBER 1-2, 2011
KECK CENTER OF THE NATIONAL ACADEMIES
WASHINGTON, D.C.

Thoughts on Estimating Long Term Science and Engineering (S&E) Workforce Needs
Timothy Coffey
National Defense University

Bureau of Labor Statistics Projections: Overview of Methods and Results
Dixie Sommers, Assistant Commissioner of Labor Statistics
Bureau of Labor Statistics

DOD Civilian Personnel Perspective on Science, Technology, Engineering, and Mathematics
Pasquale (Pat) Tamburrino, Deputy Assistant Secretary of Defense
Civilian Personnel Policy

Welcome and Purpose of Meeting
Norman R. Augustine and C. Dan Mote, Committee Co-Chairs
Committee on Science, Technology, Engineering and Mathematics (STEM) Workforce Needs for the U.S. Department of Defense and the U.S. Defense Industrial Base

From PE to EE: A Naval Postgraduate School Study of Turning Non-Technical Undergraduate Majors into Technical Program Graduates
Daniel Oliver, President
Naval Postgraduate School

Global Innovation Network and Engineering Workforce (video teleconference)
Vivek Wadhwa, *Visiting Scholar, School of Information*
University of California, Berkeley

National Academy of Engineering and National Research Council Study Committee
Emily DeRocco, President
The Manufacturing Institute

Community Colleges: Fast Facts
Ronald Williams, Committee Member
Committee on Science, Technology, Engineering and Mathematics (STEM) Workforce Needs for the U.S. Department of Defense and the U.S. Defense Industrial Base

Projects of Education Demand for the STEM Future Workforce
Nicole Smith, Senior Economist
Georgetown University

MEETING 5
MARCH 7-8, 2012
KECK CENTER OF THE NATIONAL ACADEMIES
WASHINGTON, D.C.

Closed Meeting

MEETING 6
MAY 7-8, 2012
KECK CENTER OF THE NATIONAL ACADEMIES
WASHINGTON, D.C.

Closed Meeting